Himalayan Soap Pod Tree (*Gymnocladus assamicus*): An Ecologically and Economically Important Tree on the Brink of Extinction

Himalayan Soap Pod Tree (*Gymnocladus assamicus*): An Ecologically and Economically Important Tree on the Brink of Extinction

Dr Baharul Islam Choudhury
Department of Ecology and Evolutionary Biology
University of Toronto
25 Willcocks Street, Toronto ON, M5S 3B2 (Canada)
Email: baharul.choudhury@utoronto.ca; baharulchoudhury@gmail.com

Dr Mohammed Latif Khan
Department of Botany
Dr Harisingh Gour Central University
Sagar, Madhya Pradesh (India)
Email: khanml61@gmail.com

CABI

CABI is a trading name of CAB International

CABI
Nosworthy Way
Wallingford
Oxfordshire OX10 8DE
UK

Tel: +44 (0)1491 832111
Fax: +44 (0)1491 833508
E-mail: info@cabi.org
Website: www.cabi.org

CABI
745 Atlantic Avenue
8th Floor
Boston, MA 02111
USA

Tel: +1 (617)682-9015
E-mail: cabi-nao@cabi.org

A catalogue record for this book is available from the British Library, London, UK.

ISBN-13: 978 1 78639 198 8 (hardback)
978 1 78639 199 5 (ePDF)
978 1 78639 200 8 (ePub)

Commissioning Editor: Ward Cooper
Editorial Assistant: Emma McCann
Production Editor: James Bishop

Typeset by Spi, Pondicherry, India
Printed and bound in the UK by Severn, Gloucester

Contents

Acknowledgements

This work has been accomplished with immense support and contributions from people across different walks of life and we extend our heartfelt gratitude to all. We sincerely thank the village headmen (Gaon Bura) and villagers of Dirang (West Kameng) for showing interest in our research and towards conservation of *Menangmanba-shi* or the soap pod tree (*G. assamicus*) and sharing their knowledge during this exploration. Especially, we wish to thank Mr Sang Khandu (Barchi village) and Mr Jam Tsering (Yewang Village) of Dirang who helped immensely during the field work. Mr Chanda of SFRI Orchid Orchard, Dirang, assisted in the greenhouse experiment at Dirang and we are thankful to him.

We are grateful to Dr M. Bhattacharya, (previous Director, NRC Yak, Dirang) for providing laboratory facilities, Mr Pijush Kumar Dutta, WWF-Nature, India, Dirang Field Office for providing field support, Dr Anwaruddin Choudhury for input in identification of bird species, and Principal Chief Conservator of Forest, Itanagar for logistic support during the period of work. We also acknowledge the services received from the instrumentation division, IIT Guwahati during SEM of pollen and seed samples. Help received from BSI Shillong and Itanagar and SFRI Itanagar is also appreciated. It is a pleasure to acknowledge the help received from Mr Bijit Basumatry and Mr Bironjay Basumtary during field and laboratory work.

Special thanks to Dr Ashesh K. Das of Assam University Silchar for invaluable comments on earlier versions of the chapters. Guidance received through research collaborations from Dr Shaily Menon and Dr A. Townsend Peterson is gratefully acknowledged. We also extend our sincere gratitude to Dr S. Dayanandan, Dr A. Arunachalam, Dr K. Arunachalam, Dr L.B. Singha, Dr O.P. Tripathi, Dr Kh. Ashalata Devi, Dr P. Gajurel, Dr A.R. Barbhuiya and Dr A. Paul for insightful suggestions and feedback during the entire research journey. Thanks to Madeline Jarvis-Cross for valuable comments on English language. We also thank the anonymous reviewers who read the book proposal and recommended it for publication. Financial support received from the Department of Science and Technology, Govt. of India (Sanction no. SR/SO/PS-16/2002) is duly acknowledged.

BIC is deeply indebted to the family members, especially Sabnam, Aryan and Rayan, for constant encouragement and support throughout the course of this research. MLK is thankful to family members Yasmeen Khan and Mohsneen Khan with special mention of his late wife Ms Shaheen Khan for constant support throughout life.

General Introduction

Gymnocladus assamicus is an archaic tree with a rounded spreading crown endemic to the eastern Himalayan region of northeast India. Locally known as *Menangmanbashi* by the Monpa tribe of Tawang district in Arunachal Pradesh (India), meaning 'soap pod tree', mature pods contain high saponin which is used traditionally for cleansing purposes. The soap pod tree is also intricately associated with their local tradition and religious activities. The tree remained unreported for more than 70 years in its home range and was included in a priority list for the national plant recovery programme in India. Extensive field surveys and employment of ecological niche modelling approaches could locate only 28 mature trees from 14 discrete populations, of which only 9 trees were actively reproducing. Our study thus revealed 'critically endangered' status of the species and therefore its need for a special recovery plan.

We identified both natural and anthropogenic pressures to regeneration of the tree and developed suitable strategies for an effective recovery programme. Hard seed coat imposed seed dormancy, seed hoarding by animals and over-harvesting of mature pods were found to be major limitations of regeneration for *G. assamicus*. More importantly, a very rare mating system known as 'androdioecy' was discovered, in which male and hermaphrodite individuals coexist in natural populations. The androdioecious mating system is considered an intermediate step towards the evolution of dioecy; however, it is not clear whether androdioecy in *G. assamicus* may have evolved from dioecy as a result of selection for male function in females for reproductive assurance during colonization. Phylogenetic studies in the light of mating systems among related species will be helpful in understanding evolution of androdioecy in *G. assamicus* and other such species.

Overall, the book highlights the story of a critically endangered tree species having unique biological and sociocultural importance. Our initial experiment on 'conservation through cultivation' showed great potential to recover *G. assamicus* from extinction. The book will serve as a case study and provide guidelines for conducting similar research for other rare/endangered species in the region as well as elsewhere in the world.

In Chapter 1, the biology, distribution and population status of *G. assamicus* is elaborated with informative field data. The genus *Gymnocladus* belongs to the family Leguminosae and subfamily Caesalpinioideae. It has a basal position in phylogenetic schemes and is highly diverse in floral form and ontogeny. *G. assamicus* mostly grows on hill slopes and along the banks of streams. The tree remains completely leafless for over two months during winter. New leaves appear during early March followed by flowering during April which lasts only for 15–20 days. Fully mature pods persist on the trees until the next flowering season.

An attempt has been made to understand the phylogeny and phylogeography of the genus *Gymnocladus* and its close relatives in Chapter 2. *Gymnocladus* is a small archaic genus belonging to the family Leguminosae, having very few species and showing interesting patterns of distribution. The genus was first described by Lamark with two species *Gymnocladus canadensis* and *Gymnocladus arabicus* from eastern Canada and Egypt respectively. *Guilandina dioca* (L.) is an earlier name for the Canadian species *Gymnocladus canadensis* which is again synonymised to *Gymnocladus dioica* (L.) K. Koch. Another sister genus of *Gymnocladus* is *Gleditsia* which was described by Linnaeus with a single species *Gleditsia tricanthos* from Virginia (North America). Currently, there are about 14 species in the world out of which 10 species are found in temperate Asia, 2 or 3 in North America, *G. amorphoides* (Griseb.) Taub. in South America and *G. caspica* Desf. around the Caspian Sea. In India, the genus *Gleditsia* is represented by *G. assamica* Bor. and *G. delavayi* Franch and is also restricted to the northeastern region of India. Occurrence of *Gymnocladus burmanicus* from Arunachal Pradesh in India is also highlighted in this chapter.

Chapter 3 presents a detailed account of the *G. assamicus* reproductive ecology which was revealed through pollination ecology experiments conducted in the field. Breeding systems in the genus *Gymnocladus* are confusing and described either as polygamous or dioecious or with unisexual flowers. Mature *G. assamicus* trees bear male and bisexual flowers on separate trees. Morphology of flowers in both the morphs is similar except that male inflorescences possess more flowers than hermaphrodites. Controlled breeding experiments confirmed clear sexual dimorphism with males lacking female function and hermaphrodites with both sexual functions. Such co-occurrence of male and hermaphrodite individuals in the same population is known as functional androdioecy and is very rare.

Seed biology and modes of regeneration are important criteria for management of threatened species and are highlighted in Chapter 4. Information on the ecological requirements at the seed and seedling stages is vital both for conservation and rehabilitation and is seldom articulated in silvicultural and forest management plans. The present study, based on scanning electron microscopic observation, showed the very compact nature of testa which may prevent uniform intake of water throughout the seed surface. This may have led the seed to become highly impermeable in nature. Different seed pretreatments in laboratory conditions could break the dormancy effectively and may be used in forestry practices.

Seedling dynamics of *G. assamicus* are highlighted in Chapter 5 providing results of in situ and ex situ studies. A seedling survival and growth study revealed that *G. assamicus* expresses full growth and development in rich soil moisture areas under lower irradiance levels. Therefore, plantations of *G. assamicus* should be raised along river or stream banks where soil moisture is sufficient even during drier seasons. Seedlings may be planted under medium irradiance levels along established forest/plantations with small or medium gaps.

Conservation and management of *G. assamicus* is critical. Chapter 6 provides a general biodiversity profile of northeast India and specific details for management of *G. assamicus*, highlighting both short-term and long-term goals. A conservation through cultivation approach may become successful in the case of *G. assamicus* for its brilliant autumn colour as well as its ethnobotanical importance.

Biology, Distribution and Population Status of *Gymnocladus assamicus*

1

1.1 Introduction

The eastern Himalayan region of northeast India is a global 'biodiversity hotspot', home to relict species found nowhere else in the world. However, the region's biodiversity has been threatened by anthropogenic activities such over-harvesting, habitat degradation and agricultural extension, and thus requires an integrated approach for effective management at local and international levels (Dong *et al.*, 2017). Soap pod tree (*Gymnocladus assamicus* Kanj. ex. P.C. Kanjilal) is a tree species facing tremendous extinction pressure in its natural habitat due to several such activities. The species possesses enormous economic and ecological importance throughout its distribution range and an effective conservation initiative is urgently needed. *G. assamicus* is a medium-sized deciduous tree with horizontal ascending branching. Etymologically, *Gymnocladus* means 'naked twigs' (Greek), a name which is derived from the tree's branches, which remain leafless for a prolonged period and appear 'naked' after leaf fall. *G. assamicus* is locally known as *Menangmanba-shi* among the Monpa tribe and *Minkling* among the Lish Monpa tribe in Arunachal Pradesh, meaning 'soap tree'.

Ethnobotanical use of the *G. assamicus* pods is very popular among the local people, and is intimately associated with the culture and tradition of the region. However, the species has been listed recently as critically endangered in the International Union for the Conservation of Nature (IUCN) Red List due to natural and anthropogenic activities, and requires urgent protection measures (Saha *et al.*, 2015). Natural populations of actively reproducing *G. assamicus* trees were found to be extremely small (Menon *et al.*, 2010) and, as such, the species has been included on the priority list for the recovery programme in India (Ganeshaiah, 2005).

1.2 Taxonomy of *G. assamicus*

The taxonomic position of *Gymnocladus assamicus* as par Benthum and Hooker (1862–1883) is as follows:

- **Phylum:** *Angiospermae*
    - **Class:** *Dicotyledonae*
        - **Subclass:** *Polypetalae*
            - **Order:** *Rosales*
                - **Family:** *Leguminosae*
                    - **Subfamily:** *Caesalpinioideae*
                        - **Genus:** *Gymnocladus* Lamarck, 1785
                            - **Species:** *G. assamicus* Kanjilal ex. PC Kanjilal

Botanical name: *Gymnocladus assamicus* Kanjilal ex. PC Kanjilal, 1934.

Type: India, Assam, Khasi Hills, Laitkshe, Marngor, 1800 m, UN Kanjilal 7624, holo: DD; iso: ASSAM, CSL, K.

1.2.1 Taxonomic description

This is a medium-sized tree with average height of 15 to 17 m (Fig. 1.1.); the trunk is horizontal with ascending branching. The bark is greyish-brown with an outer reticulate corky layer having parallel furrows (Fig. 1.2). Bi-pinnate compound leaves are alternate, 38 15 x 20–25 cm, obscurely glandulose with a swollen base, having 10–12 pairs of pinnae (Fig. 1.3). The pinnae are opposite or distantly sub-opposite, 10–22 x 6–8 cm with 15–20 leaflets. Leaflets are shortly petioled, subopposite to alternate, 2–2.3 x 0.64–0.67 cm, oblong to ovate oblong, minutely mucronate, nearly glabrous above with minutely brown pubescent beneath the midrib. The main lateral nerves are 5–8 in number on either side and are very obscure, slightly elevated above and depressed beneath. The base of the leaflets is rounded to slightly cuneate with short petiolules mounted with sparsely adpressed hair.

1.2.2 Floral sexuality

The inflorescence of *Gymnocladus assamicus* is of terminal racemose type, having fine pubescence throughout. Male inflorescence is larger, 13–16 x 5–6 cm, with nearly whorled lateral branches comprising 15–20 flowering nodes (Fig. 1.4). In comparison, hermaphrodite inflorescence is smaller and varies from 4–6 x 5–6 cm (Fig. 1.5). The average number of flowers per male inflorescence was observed to be 72.50 (±4.01, n=20) and that of hermaphrodite flowers per inflorescence was observed to be 23.40 (±4.01 n=20). Both male and bisexual flowers are small, tubular, violet-coloured and odourless.

Fig. 1.1. Habitat of a mature *G. assamicus* tree (photo: Baharul I. Choudhury).

Fig. 1.2. *G. assamicus* bark: reticulate corky outer cover having parallel furrows (photo: Baharul I. Choudhury).

Fig. 1.3. Bi-pinnate compound leaves of *G. assamicus* (photo: Baharul I. Choudhury).

1.2.3 Hermaphrodite flower

Complete, pedicellate, actinomorphic, pentamerous, tubular, narrowed at the base and becoming gradually wider at the mouth. Corolla is with five united petals. Petals 10–15 mm long, hairy, shaggy throughout, corolla tube ca 3.43 mm wide. Calyx with five basally connate sepals, subequal, lanceolate to ovate, 4–6 mm long with shaggy brown hair. Corolla tube is 5–7 mm long and 3.43 mm wide at anthesis and is also shaggy brown hairy outside.

Calyx comprising five lobes, which are subequal and lanceolate to subulate, 4–6 mm long with shaggy brown hairy outside. Stamens 10, didynamous, alternate with 1.5–2.5 mm long filaments. Anthers are oblong, ovoid and dithacous.

Fig. 1.4. Male *G. assamicus* inflorescence (photo: Prabhat C. Nath).

Fig. 1.5. Hermaphrodite *G. assamicus* flowers (photo: Baharul I. Choudhury).

Pistil one, 9–10 mm long at anthesis, slightly arched, thick and compressed, well demarcated into stigma, style and ovary. Ovary superior, unilocular with 5–8 ovules in each locule, style straight, slightly thick and compressed with oblique stigma. Stigma surface is slightly slanted, papillate with wet sticky exudates.

1.2.4 Male flower

Male flowers are shorter pedicellate as compared with hermaphrodite flowers. Each corolla tube is 4–6 mm long and ca 3 mm wide at anthesis and is also shaggy brown hairy on the outside. Vestigial carpel is visible at the base of the perianth tube.

1.2.5 Fruit

The fruits of *G. assamicus* are of typical pod type and range from 10–16 to 2.5–4 cm in dimension, and are compressed and turgid over the seeds. Pericarp is polished and fleshy; mesocarp is highly saponaceous (Fig. 1.6). Each pod contains 4–8 seeds. Seeds are 14–15 x 15–17 mm in dimension, ovoid or subglobose, have black horny testa, and are extremely hard. Radicle is comparatively small and erect.

Fig. 1.6. Highly saponaceous *G. assamicus* mature pods (photo: Baharul I. Choudhury).

Fig. 1.7. *G. assamicus* mature pods from previous fruiting season (photo: Baharul I. Choudhury).

Fully mature pods are fleshy, having a pungent smell which persists on the trees for a prolonged period of time (Fig. 1.7). Wood is moderately hard and yellowish white. Pith is soft, spongy and more prominent in young shoot (Fig. 1.8).

1.3 Ethnobotanical Uses of Fruits

Ethnobotanical information forms a rich knowledge system for the health and nutrition of indigenous people, particularly within resource-poor rural communities.

Fig. 1.8. Longitudinal section of *G. assamicus* wood (Photo: Baharul I. Choudhury).

Such invaluable information is often unwritten and inherited through generations via verbal communication. However, this knowledge system is being threatened by modern cultural practices and global economic changes. Therefore, documentation and preservation of ethnobotanical knowledge is crucial for sustainability of natural resources, preservation of tradition, resource management, conservation and community development.

Traditional societies in the eastern Himalayan region are integral to the local ecosystems, and play a significant role in the maintenance of their ecological surroundings. Non-timber forest products (NTFPs) have been collected and used for generations for subsistence and trade. *G. assamicus* pods are unique in multiple aspects. Pods become brown and fleshy during maturation and develop an alluring smell. Post-maturation, pods persist on trees for an extended period and are clearly visible (Fig. 1.7). This trait helps local people locate mature pods from a fair distance and harvest the pods for future use.

The local Monpa people are largely dependent on NTFPs for food, medicine, dye, firewood, resin and fibre (Saha and Sundriyal, 2012). *G. assamicus* is intimately associated with the daily life and rituals of the Monpas of Arunachal Pradesh due to its multiplicity of ethnobotanical uses (Choudhury *et al.*, 2007a). Local people are familiar with the locations of most of the mature trees in the neighbourhood, and harvest fully mature pods mainly from the tree directly, or from the ground. Large-scale harvesting, often of all mature pods, has been found to be common. Excess pods are stored for future use, sometimes for four to five years, as documented in the Monpa community. In some cases, pods are harvested by cutting down the branches, which severely damages the tree (Fig. 1.9). Monpa people consider mature *G. assamicus* pods sacred, and exchange them as gifts within the community. Additionally, Buddhist monks, locally known as *lama*, gratefully accept the pods as religiously significant gifts.

Pods are exploited for multiple purposes. They are primarily used during worship at monasteries, locally known as *gompa*. The Buddhist people of Arunachal Pradesh prepare a decorative, colourful and sacred structure known as a *torgen*, meaning 'flower' (Fig. 1.10), to perform rituals. The main ingredient of the *torgen* is *torma* which is a mixture of wheat flour, clarified butter (ghee) from yak milk, and wax. *Torma* is very sticky in nature, and clings to the hands during *torgen* preparation. Pre-soaked pods are rubbed between the hands to prevent such clinging during preparation, as a substitute for detergent (Fig. 1.11). The use of *G. assamicus*

Fig. 1.9. Mature pods have been collected here by cutting down branches of the *G. assamicus* tree (photo: Baharul I. Choudhury).

pods is more effective than commercial soaps in eliminating the stickiness of the *torma*. More importantly, use of this natural soap substitute for such rituals is regarded as sacrosanct.

Cleanliness is a human necessity, and traditional ways of maintaining cleanliness using available means and methods have been practiced for centuries. Multiple plant species with rich saponin content and effective cleansing properties have been reported in India (Mehta and Bhatt, 2007) and other parts of the world (Osbourn, 1996). The fruit of *Sapindus mukorossi*, popularly known as soapnut, is recognized in many cultures across Asia for its cleansing properties (Nakayama *et al.*, 1986; Suhagia *et al.*, 2011). The cleansing action of soapwort (*Saponaria officinalis*) has also been known for centuries (Osbourn, 1996). However, due to a lack of documentation, *G. assamicus* pods are lesser known despite their strong cleansing properties. Pods are popularly used for domestic cleansing purposes such as washing clothes, hands and utensils. Older generations still prefer the pods for bathing, in lieu of soap. We documented a very efficient cleansing properties of the pods against a wide range of strong fats, oils and other complex stains from daily activities. Women use the pods for shampooing, and cite them as superior to commercial shampoos for the removal of dandruff. In one study, MeOH extracts from the fruits of *Gymnocladus* were found to exhibit anti-HIV activity (Konoshima *et al.*, 1995). In another study, Lee and Morris-Natschke (1999) found that *Gymnocladus* saponin isolated from fruits inhibited HIV replication with an EC_{50} value of 27μM. Dried fruits of *G. chinensis* have long been known

Fig. 1.10. *Torgen* made of highly greasy materials assisted with the use of *G. assamicus* pod (photo: Baharul I. Choudhury).

Fig. 1.11. Demonstration of *G. assamicus* pod use during *torgen* preparation in Dirang monastery (photo: Baharul I. Choudhury).

in oriental medicine as saponin drugs, and used as diuretics and expectorants (Konoshima *et al.*, 1995). A recent study reported the synthesis of gold (Au) nanoparticles by using *Gymnocladus assamicus* pod extract in an aqueous medium (Tamuly *et al.*, 2013). However, chemical properties of the saponin are not fully known or characterized chemically as yet, and this may provide a better understanding of the active compounds in the mature pods.

Forest dwellers in Tawang and West Kameng districts of Arunachal Pradesh use the decoction of pods on the bodies of their domestic animals to prevent leech and other insect bites. Pods are soaked overnight in water and applied to the animals externally before they are let out to graze in forest land. Black-coloured seeds are round and extremely hard. Local people occasionally munch roasted *G. assamicus* seeds, or prepare beverages by boiling the powdered seeds.

The North American *G. dioca* is commonly known as the Kentucky coffee tree (USDA, 2017), a name which stems from the resemblance of its seeds to coffee beans. The early immigrants to Kentucky are said to have roasted and ground the coffee-tree beans, and used them as an inferior substitute for coffee. There are some reports that indigenous North Americans used the seeds for food. However, the seeds are unpleasantly bitter, and contain the alkaloid cystosine, which causes gastrointestinal disorders and may lead to irregular pulse and coma (Bowles, 2004). Overconsumption of *G. assamicus* seeds causes dizziness, nausea and vomiting. Cattle fatalities have also been reported after ingesting the mature seeds.

The dark, round, shiny seeds also have some aesthetic appeal and are used as beads in rustic jewellery. *G. dioca* seeds are known as 'hully-gullies' and are sometimes carried by children as good luck charms. The wood of the tree is coarse-grained, heavy and strong. In the USA, it is sometimes used for railway ties, fence-posts and construction but is of little commercial importance due to its lack of abundance. *G. dioca* is also grown extensively as an ornamental tree (Bowles, 2004).

1.4 Distribution

The edapho-climatic conditions of northeastern India favour luxuriant growth of vegetation attaining the stage of climax formation which is represented by the scattered oak forests of the region. Arunachal Pradesh is located in the extreme northeastern part of India with a geographical area of 8.37 million ha and is the largest state in the region. It consists of 16 districts and is home to extraordinarily high levels of biodiversity (Dong *et al.*, 2017). The state has a wide range of elevations, ranging from 100 m in adjacent areas of the Assam plains, to 7000 m in the Shiwalik hill ranges of the greater Himalayas. The species richness of the state is very high, with 4,000–5,000 species of vascular plants per 10,000 sq. km, and it is thus considered the second most biodiverse region on the globe (Valdiya, 2002). It is endowed with six major forest types within four major climatic categories (Kaul and Haridasan, 1987) and is home to 4,117 species of flowering plants belonging to 1,295 genera and 192 families (Chowdhery and Pal, 1997). Major groups include 600 species of orchids, 89 species of bamboos, 18 species of canes, 400 species of ferns, 24 species of gymnosperms and innumerable species of algae, fungi, lichens, bryophytes and microorganisms (Hegde, 2002). New taxons are routinely added from the area (Rappole *et al.*, 2005; Bawri *et al.*, 2016) which harbours great potential for continued discovery through field exploration and the use of geographic information system (GIS) technologies.

G. assamicus is rarely documented due to its extremely small natural population. After its first report from the Khasi Hills of Meghalaya (undivided Assam) in 1934 (Kanjilal *et al.*, 1938), the species remained undocumented for several decades except two sporadic records from Arunachal Pradesh (Hajra *et al.*, 1996) and Nagaland (Sanjappa, 2000) of northeast India. It has been re-documented recently from the West Khasi Hills district of Meghalaya (Venugopal and Pamidimarri, 2015). We performed an extensive field survey throughout Arunachal Pradesh to document the distribution of *G. assamicus*, and recorded individual trees from each site from 2004 to 2007.

Locating populations of such rare, endangered and poorly understood species is an important consideration in biodiversity conservation for assessing species' conservation status and for developing effective strategies for in situ and ex situ conservation. Poor records of *G. assamicus* populations may be associated with challenging field surveys constrained by rugged terrain, political unrest, poorly understood distributional ecology and/or low population sizes with widely dispersed individuals. To overcome these challenges, we applied an ecological niche modelling (ENM) technique which provides a powerful predictive framework for targeting areas for field surveys to locate additional populations. ENM combines known occurrence records with relevant environmental layers to estimate species' ecological requirements and potential geographic distributions (Guisan and Zimmermann, 2000; Soberón and Peterson, 2005). The technique assists in narrowing down the most plausible occurrence sites for targeted field surveys, which otherwise may constitute a formidable time and labour sink with low probability of success. The ENM approach has been successfully used in locating new populations of poorly understood species (Raxworthy *et al.*, 2003; Bourg *et al.*, 2005; Menon *et al.*, 2012; Jathar *et al.*, 2015).

We focused on Tawang and West Kameng districts in Arunachal Pradesh for this study (26° 54′ to 28° 01′ N, 91° 16′ to 92° 40′ E). The region is bordered by Tibet to the north, Bhutan to the southwest and the state of Assam to the south. Detailed field inventories were carried out between 2004 and 2007 in the Dirang forest area located in West Kameng, Arunachal Pradesh, northeastern India (Fig. 1.12). The forest is variously classified as eastern Himalayan wet temperate forest (Champion and Seth, 1968) or subtropical deciduous forest (Kaul and Haridasan, 1987). The area is mostly mountainous, covering 892 km^2, and is surrounded by higher elevation terrain (2,800–3,400 m) in Bomdila and Tawang districts. The Dirang Forest Range was declared an 'unclassed state forest' by the state forest department as most of the area is under the jurisdiction of local communities. The forest areas are exposed to diverse forms of anthropogenic and natural influences, including logging, agricultural extension, human settlement, road construction and landslides. Detailed field inventories of natural populations of *G. assamicus* were collected from local inhabitants and direct field visits to nine locations. The type locality for *G. assamicus* is located approximately 200 km southwest of the present location in Latikesh, Marngor, Assam, Khasi Hills, now

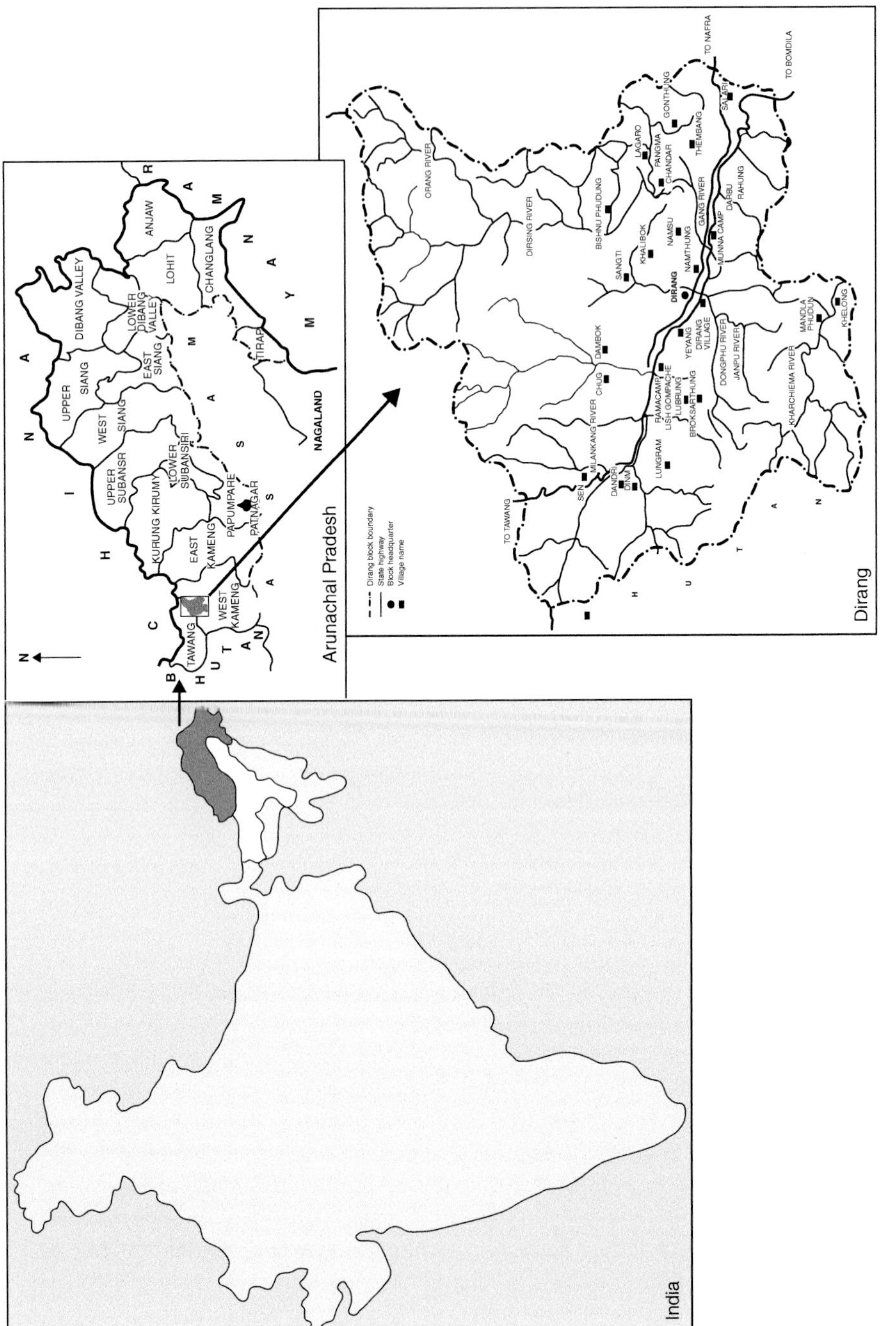

Fig. 1.12. Location map of Dirang showing distribution of *G. assamicus* populations (starred numbers).

in Meghalaya. However, it may be extirpated from the site since no collection has been made since 1920. All occurrence locations were georeferenced using geographic/global positioning system (GPS) units accurate to 10–100 m. Finally, we recorded 14 discrete populations of one to seven trees each and the occurrence points were used as input data to train the niche model. We then applied two ENM techniques (GARP and Maxent), along with a jackknife validation approach, to develop ENMs based on the few known occurrences.

Eleven remotely-sensed layers from the NASA-MODIS/Terra data set (spatial resolution 500 x 500 m – Justice *et al.*, 1998) were used to characterize environments across the region. Six of the layers were 16-day composite images of the enhanced vegetation index (EVI) from every second month in 2005, and five summarized the difference between each consecutive pair of the six EVI layers. Two evolutionary-computing approaches were used to develop ENM and associated geographic predictions: the genetic algorithm for rule-set prediction (GARP) (Stockwell, 1999) and Maxent (Phillips *et al.*, 2006). Both algorithms relate known occurrence points to sets of digital raster GIS data layers that describe relevant environmental parameters (in this case, the EVI data sets) to derive a model in the form of decision rules differentiating areas of suitable conditions from areas presenting unsuitable conditions. GARP uses a genetic algorithm to optimize the predictive ability of models regarding independent suites of occurrence information, whereas Maxent uses the principle of maximum entropy to achieve the same end. In this application, we used default parameterization of both algorithms (except that a soft omission threshold of 20% was used in GARP, and a 50% random test percentage was used in Maxent). Both algorithms are well described in the literature and have been used to predict species occurrences with maximum success in locating the species of interest (Stockwell, 1999; Peterson *et al.*, 2007; Batalden *et al.*, 2014; Raghavan *et al.*, 2016). The ENM approach was designed, developed and implemented in collaboration with Dr Shaily Menon from Grand Valley State University (USA) and A. Townsend Peterson from the University of Kansas (USA). Detailed ENM methodology is described in Menon *et al.* (2010).

Altogether, 26 discrete areas were predicted by the ENM as having the highest probability of occurrence (i.e. most highly suitable) for *G. assamicus*. Maps were generated to guide the field survey demarcating the centre and search radius for each of the 26 areas. The search radii ranged from 0.5–2.0 km, corresponding to the extent of the high probability prediction at each site. A rapid field survey was conducted over a 10-day period with the help of the maps and a GPS. Of the 26 predicted areas, 14 were relatively accessible. The field survey team directly visited each of these 14 sites and searched for *G. assamicus* individuals within the radius predicted by the niche model. The remaining 12 areas could not be visited due to inaccessibility. Instead, the field survey team interviewed residents from nearby villages to gather indirect evidence.

The ENM prediction from GARP and Maxent identified small patches of suitable areas across the region (Fig. 1.13) and the average of the two niche modelling approaches was used to predict the potential distribution of *G. assamicus*. The

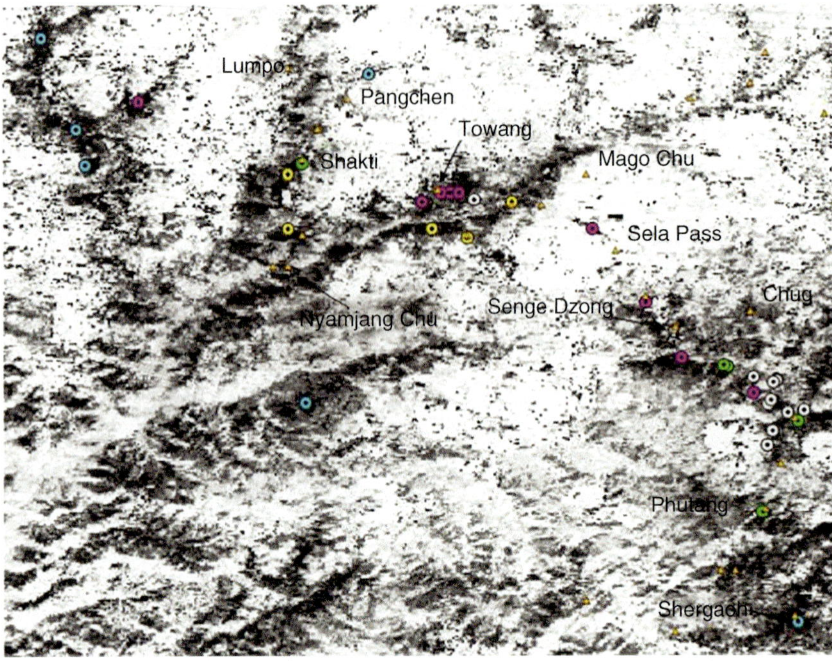

Fig. 1.13. ENM map showing *G. assamicus* populations in the mapped area. Darker patches represent regions with higher probability of the species' occurrence. White circles represent input data (14 proviously known occurrence points) used to create the models; green circles represent direct evidence of the species' presence; yellow circles represent indirect evidence of the species' presence; purple circles represent the species' absence; and blue circles represent areas not yet surveyed.

predictions also identified similar areas across the study area. The jackknife tests of model validations confirmed that both predictions were predicting the species' occurrence significantly better than random expectations ($p<0.05$). Averaging the two predictions produced a predictive map that identified relatively restricted portions of the study areas as presenting the highest suitability for the species (Fig. 1.13). The field surveys encountered five new occurrences of *G. assamicus* at 14 of the 26 suitable sites that could be directly surveyed.

Populations comprising four mature trees each were encountered in Sapper Camps 1 and 2 at the Chug predicted site. Both populations were on slopes on the bank of a small river. Although the trees were mature, flowering and fruiting were not observed during the field survey. Two fruiting trees were found growing between small hillocks in the Dirang Dzong (Phadam) predicted site. One mature tree was discovered at each of the Shakti and Phudung (Wangkethanka) sites in valleys near small streams. Interviews of villagers from nearby villages indicated the likely presence of additional trees at five sites that were not accessible for field visits.

This study illustrates the potential of ENM in enabling and improving endangered species research, permitting identification of additional populations of rare and poorly understood species via rapid field surveys, targeted and guided by niche model predictions. Our study demonstrates that models generated with small numbers of known occurrence points can be effective in predicting high probability areas of species occurrence. This validation approach has demonstrated the considerable predictive power of models generated with scarce locality records with respect to their potential for the discovery of new species and populations (Peterson *et al.*, 2007). More detailed field surveys, including areas where only indirect evidence of the species' presence was recorded, will likely uncover additional populations and individuals of this species. Despite our discoveries, the numbers and population size of *G. assamicus* remain extremely small.

1.5 Population Status of Soap Pod Tree and its Associated Vegetation

The population dynamics of plant species can be described by demographic variables such as the recruitment, mortality and growth rates of individuals (Watkinson, 1997). Patterns of birth and death determine population size and vary drastically within species (Roff, 1993). Balance among these variables has been found to regulate the dynamics and structure of a population (Kohyama and Hara, 1989). The structure of plant communities in any forest ecosystem is largely determined by intensity, magnitude and frequency of disturbances that take place in a forest (Armesto and Pickett, 1985; Miller *et al.*, 2011). Disturbance often causes different effects on vegetation including canopy opening and creating treefall gaps that provide opportunities for the quick appearance of pioneer tree species in areas where direct sunlight reaches the ground (Brokaw, 1982; Whitmore, 1984; Holeksa *et al.*, 2017). On the other hand, intense disturbance events may destroy a larger section of the forest, leading to more widespread regeneration (Wright and Heinselman, 1973). Quantitative studies have shown that pioneer species have fast growth rates, short life spans, low wood density and efficient seed dispersal (Swaine and Hall, 1983; Whitmore, 1984). Different types of disturbances differentially affect vegetation (Segura *et al.*, 2003). For example, disturbance may increase biodiversity within a community (Thom and Seidl, 2016), thus enacting positive change. In contrast, disturbance may also disrupt the climax species, thereby destabilizing the community (Clements, 1936) and decreasing the density and basal area of tree species (Nath *et al.*, 2005), thus enacting negative change.

A few plant species are restricted to very small areas due to specific microclimatic conditions (Wulff *et al.*, 2013) or climatic regression after the last Ice Age (Cox and Moore, 1993; Garcia *et al.*, 2000). Due to climate change and human disturbance, distribution of many species has reduced considerably or has become limited to small isolated patches with fewer number of individuals or migration of species towards suitable habitats (Franco, 1986; Garcia *et al.*, 2000; Corlett and Westcott, 2013). Species of such limited distribution often face acute regeneration failure, leading to extinction. Many authors have studied the regeneration status of tree species in various parts of the world and have predicted regeneration status based on the age and size structure of a population (Khan *et al.*, 1987; Bhuyan *et al.*, 2003; Götmark and Kiffer, 2014). Successful natural regeneration is dependent upon several intrinsic and extrinsic factors such as the production and germination of seeds, sufficient seedling production, and establishment of seedlings and saplings (Good and Good, 1972; Rao, 1988). Biotic and abiotic environmental factors also greatly influence natural regeneration (Khan and Tripathi, 1986; Hirayama and Sakimoto, 2003, Tingstad *et al.*, 2015). However, very few studies have been conducted on population structure and regeneration status of rare/endangered northeast Indian species. The present study was undertaken to understand the population structure and regeneration status of *G. assamicus* in three different sites: Changfu Moon, Dambla village and Moishing village of the West Kameng district, Arunachal Pradesh.

Vegetation associated with *G. assamicus* was studied from 2004 to 2006 to understand the community structure by quadrat method. Multiple study plots were selected after preliminary field surveys based on the availability of *G. assamicus* population and geographic accessibility. Nine different plots were earmarked where *G. assamicus* is present, out of which three representative sites, that is, Changfu Moon, Dambla village and Moishing village, were selected for vegetation and microclimate analysis. Dambla village and Moishing village sites are located near to small streams and have gentle slopes (25°–30°), whereas Changfu Moon is located on a hilltop with a steep slope (50°–60°). Physical conditions of the hill slopes are drier compared to the stream sites. Geography, altitude and aspects of each of the study sites are given in Table 1.1. Distance between and among the sites ranges from 3 km (between Changfu Moon and Dambla village) to 8 km (between Moishing village and Dambla village).

To record the tree species, 20 quadrats of 10 m x 10 m size (covering 0.2 ha area) were randomly laid in the three selected sites. Similarly, 20 quadrats of 5 m x 5 m for shrubs and 20 quadrats of 1 m x 1 m for herbs were laid randomly within the same above-mentioned 10 m x 10 m quadrats in each of the three sites.

Girth at breast height (gbh) for trees and collar diameter (cd) for saplings (10≥20 cm cd at base and >30 cm height) and seedlings (≤10 cm cd at base and <30 cm height) were recorded. Seedling population and survival in relation to distance from the mother tree were ascertained by counting the number of seedlings and saplings with respect to increase in radial diameter following Khan and

Table 1.1. Geographic location and microclimate of the three study sites.

Study site	Latitude (N)	Longitude (E)	Mean altitude (m)	Aspect	PAR (mmol s^{-1} m^{-2})
Changfu Moon	27°20′91″	92°14′16″	1922	Northwest	242–250
Dambla Village	27°20′51″	92°16′57″	1715	Southeast	1123–1130
Moishing	27°18′16″	92°14′11″	2052	Northwest	38–45

Tripathi (1986). Periodic determination of densities of seedlings was studied in one-year intervals between 2005 and 2007 in 20 randomly laid permanent quadrats of 1 m x 1 m size. All the seedlings and saplings that were recorded in each permanent quadrat were labelled with permanent aluminium tags. Survival, mortality and new recruits were estimated from such data, and demography of the species was drawn accordingly.

Plant specimens were collected and identified by consulting different herbaria (Botanical Survey of India, Itanagar and Shillong; State Forest Research Institute, Itanagar), following different floras (Kanjilal *et al.*, 1938; Hajra *et al.*, 1996; Polunin and Stainton, 2006; Stainton, 2007) and by consulting experts of the region. Nomenclature of plant species followed Benthum and Hooker (1862–1883). Quantitative analysis of vegetation for frequency, density and relative density was done following Misra (1968). Importance value index (IVI) was computed by summing up relative frequency, relative density and relative basal area for the woody species, while the same was calculated by summing up relative frequency and relative density values for herbaceous species.

The Sorenson similarity index was calculated using the formula given by Sorenson (1948).

The Sorenson similarity index = $\frac{2C}{A+B}X100$ – where C is the number of species common to two relevés, A is the total number of species in relevé A, and B is the total number of species in relevé B.

In the present study, altogether 28 mature *G. assamicus* trees were documented from nine sites located in Dirang circle of West Kameng District, Arunachal Pradesh. The sites were scattered among 10 km^2 (altitude 1500–2000 m asl) of landscape, fragmented by heterogeneous geographical barriers such as rivers, small mountains, roads, human settlements and shifting agricultural land. Each of the sites is located at a different distance from human settlements, and as a result experiences different degrees of human interference. Moishing is the furthest location from human settlements (about 5 km) and thus experiences the least anthropogenic pressure. Other sites are located in the vicinity of small villages and thus harvesting pressure is comparatively high. The species is mostly distributed on the hill slopes and along the banks of streams. Number of mature trees per site ranged from one to seven, and only two sites (Moishing and Dambla) had seedlings and saplings. The distribution, sexuality and current population status of *G. assamicus* in and around Dirang of Arunachal Pradesh is presented in Table 1.2.

A total of 60 plant species (16 tree species, 16 shrubs and 28 herbs) belonging to 37 families and 60 genera were found to be associated with *G. assamicus* in the three study sites. *Quercus griffithii* and *Rhus wallichii* were the most common trees growing with *G. assamicus* in the three studied sites. Other associated woody species were *Alnus nepalensis*, *Illicium griffithii*, *Juglans regia*, *Rhododendron arboreum* and so forth (Table 1.3). The most common herbaceous species among the three sites were *Crassocephalum crepdioides* and *Smilacina purpurea*, whereas *Artemisia indica*, *Globba clarkei*, *Hydrocotyle asiatica*, *Polygonum plebeium* and *Spilanthes paniculata* were also common among two of the three studied sites (Table 1.3).

Table 1.2. Population status of *Gymnocladus assamicus* in and around Dirang of Arunachal Pradesh.

Site	Locality	Date of survey	Altitude (m)	Latitude (N)	Longitude (E)	No. of individuals	
						M	H
1	Changfu Moon	21.09.2005	1960	27°21′	92°14′	4	–
2	Dambla Basti	02.06.2005	1723	27°23′	92°14′	2	5
3	Dirang Basti	21.09.2005	1715	27°20′	92°15′	–	1
4	Jyotinagar	8.01.2005	1724	27°22′	92°14′	1	–
5	Lishpa Village I	26.04.2007	1870	27°23′	92°13′	5?	–
6	Lishpa Village II	27.04.2007	1900	27°23′	92°13′	5?	–
7	Moishing	19.11.2004	2052	27°18′	92°14′	1	1
8	Runkung Village	04.12.2006	1669	27°22′	92°15′	–	1
9	Yewang Village	19.11.2004	1688	27°21′	92°14′	1	1

M = male; H = hermaphrodite.

Table 1.3. Density (plants ha^{-1}) and IVIs of tree species and relative importance of shrub and herb species occurring in the three selected sites.

			Changfu Moon		Dambla village		Moishing village	
Species	Family	Habit	Density ha^{-1}	IVI	Density ha^{-1}	IVI	Density ha^{-1}	IVI
Alnus nepalensis D. Don.	Betulaceae	T	1	11.21	1	12.30	–	–
Alstonia neriifolia D. Don	Apocyanaceae	T	1	6.87	0.25	2.70	4.25	26.52
Erythrina stricta Roxb.	Leguminosae	T	–	–	10.25	94.55	–	–
***Gymnocladus assamicus* Kanj. ex PC Kanj.**	**Leguminosae**	**T**	**1**	**17.31**	**1**	**9.05**	**0.5**	**7.70**
Ilex fragilis Hook. f.	Aquifoliaceae	T	1.75	11.92	1.25	11.02	–	
Illicium griffithi Hook. f. & Thomson	Illiciaceae	T	2.75	11.81	–	–	1	10.28
Juglans regia L.	Juglandaceae	T	3.5	32.02	4.75	45.11	–	–
Phyllostachys mannii Gamble	Poaceae	T	–	–	2.5	10.14	17.5	63.15
Pinus roxburghii Sargent.	Pinaceae	T	4.5	38.11	–	–	–	
Pyrus phasia Buch. Ham. ex D. Don	Rosaceae	T	1.25	8.77	0.75	8.14	–	–
Quercus griffithii Hook. f. & Thomson ex Miq.	Fagaceae	T	4.25	36.70	2.75	34.20	12.75	182.45
Rhododendron arboreum Smith.	Ericaceae	T	3.75	29.98	–	–	0.25	4.57
Rhus wallichii Hook. f.	Anacardiaceae	T	3.25	26.72	1	9.09	0.25	5.33
Rinorea bengalensis Wallich	Violaceae	T	1.25	7.67	0.5	5.47	–	–
Schima khasiana Dyer.	Theaceae	T	5.25	60.90	4.5	58.22	–	–
Achyranthes aspera Hook. F.	Amaranthaceae	H	3,950	21.7	–	–	243	11.9

Continued

Table 1.3. Continued.

			Changfu Moon		Dambla village		Moishing village	
Species	Family	Habit	Density ha^{-1}	IVI	Density ha^{-1}	IVI	Density ha^{-1}	IVI
Alstonia neriifolia D. Don	Apocyanaceae	S	600	5.9	–	–	–	–
Anaphalis busua DC.	Asteraceae	H	1,100	7.4	–	–	–	–
Arisaema consanguineum Schott.	Araceae	H	1,500	10.7	–	–	–	–
Artemisia indica Willd	Asteraceae	H	–	–	2,350	8.6	105	7.4
Betula elnoides Buch.-Ham. ex D. Don	Betulaceae	S	–	–	–	–	160	7.5
Bidens pilosa L.	Asteraceae	H	4,000	17.9	–	–	–	–
Boehmeria platyphylla D. Don	Urticaceae	S	–	–	8,300	21.5	–	–
Cirsium verutum (D.Don) Sprengel	Asteraceae	H	–	–	1,100	5.9	–	–
Clematis acuminata Bruehl	Rananculaceae	H	–	–	500	4.3	–	–
Commelina paludosa Blume	Commelinaceae	H	–	–	–	–	458	18.9
Crassocephalum crepdioides S. Moore	Asteraceae	H	750	5.7	2,550	8.9	93	7.1
Cynotis voga (Lour) Schultes & Schultes	Commelinaceae	H	–	–	4,500	13.9	–	–
Desmodium oblongum Wallich ex Benth.	Leguminosae	S	–	–	1,150	4.8	–	–
Elaeagnus caudata Schlecht	Elaegnaceae	S	–	–	–	–	53	4.9
Elsholtzia blanda Benth.	Nyctaginaceae	S	1,300	8.7	–	–	–	–
Fagopyrum esculantum Monech	Polygonaceae	H	450	4.7	–	–	180	11.6

Ficus altissima Blume.	Moraceae	S	–	–	900	3.2	–	–
Fragaria nubicola Lindley ex. Lacaita	Rosaceae	H	–	–	–	–	305	14.0
Globba clarkei Baker	Zingiberaceae	H	3,350	19.1	2,150	7.1	–	–
Glycine max L.	Papilionaceae	H	1,700	12.6			–	–
Goldfussia nutans Nees	Acanthaceae	H	–	–	1,200	4.3	–	–
Hemiphragma heterophyllum Wallich	Scrophulariaceae	H	–	–	1,100	4.7	–	–
Hydrocotyle asiatica L.	Apiaceae	H	3,550	18.4	4,450	13.3	–	–
Ilex fragilis Hook. f.	Aquifoliaceae	T	–	–	–	–	383	15.3
Ligustrum indicum (Lour.) Merr.	Oleaceae	S	–	–	–	–	263	13.6
Litsea cubeba (Lour.) Pers.	Ericaceae	S			1,400	4.7		
Lygodium japonicum (Thunb. ex Murr.) Sw.	Lygodiaceae	H	1,450	8.5	–	–	–	–
Lyonia ovalifolia C.B. Clarke) Ridley	Ericaceae	T	–	–	–	–	50	2.4
Melastoma normale D. Don	Melastomaceae	S	–	–	500	3.7	–	–
Morus macroura Miq.	Moraceae	T	–	–	–	–	283	13.5
Nepeta longibracteata Benth	Lamiaceae	H	1,000	7.8	–	–	–	–
Oenanthe javanica (Blume) DC	Apiaceae	H	–	–	–	–	95	6.5
Panicum insicum L.	Poaceae	H	–	–	4,100	11.5	–	–
Phaseolus vulgaris L.	Papilionaceae	S	1,200	7.7	–	–	–	–
Pilea umbrosa Blume.	Urticaceae	S	1,650	11.8	–	–	–	–
Polygonum plebeium R. Br.	Polygonaceae	H	–	–	1,950	6.2	113	7.6
Pouzolzia fuscescens Kuntze	Utriculaceae	H	–	–	–	–	295	13.8

Continued

Table 1.3. Continued.

			Changfu Moon		Dambla village		Moishing village	
Species	Family	Habit	Density ha^{-1}	IVI	Density ha^{-1}	IVI	Density ha^{-1}	IVI
Pyrus phasia Buch. Ham. ex D. Don	Rosaceae	S	900	9.5	650	2.8	–	–
Rubia manjith Roxb. Ex Flem.	Rubiaceae	H	–	–	6,450	21.8	–	–
Schima khasiana Dyer.	Theaceae	T	–	–	2,550	7.8	–	–
Senecio wallichii DC.	Asteraceae	H	1,900	12.6	–	–	–	–
Smilacina purpurea Wallich.	Liliaceae	H	1,450	9.2	2,350	9.1	418	16.8
Solanum torvum Swartz	Solanaceae	S	–	–	2,500	8.8	–	–
Spilanthes paniculata Wall. ex DC	Asteraceae	H	–	–	750	3.0	430	17.7
Spiraea micrantha Hook	Rosaceae	S	–	–	–	–	193	9.5
Thalictrum alpinum L.	Rananculaceae	H	–	–	1,850	6.6	–	–
Viburnum erubescens Wallich ex DC	Sambucace	S	–	–	2,900	10.1	–	–

IVI: importance value index; T = tree; H= herb; S= shrub; '–' = absence.

The overall species richness varied among sites, and the highest number of associated plant species (34) was recorded from the Dambla site, and the lowest (24) from the Moishing site. Out of the total 60 plant species, 18 were woody species and 42 were shrubby/herbaceous species. Thirteen woody and 16 herbaceous species were documented in the Changfu Moon site. Twelve woody species were found in the Dambla site, while seven woody species were found in the Moishing site. The number of herb/shrub species found was 25 and 17 in the Dambla and Moishing sites, respectively (Table 1.3). The highest number of woody species were found in the Changfu Moon site (12), while the highest number of herbaceous species were found in the Dambla site (25).

The highest stand density of woody species (36.5 stems ha^{-1}) was recorded in the Moishing site and the lowest (30.5 stems ha^{-1}) in the Dambla site. The highest basal area of woody species was recorded in the Changfu Moon site (938.19 m^2 ha^{-1}) followed by the Dambla (808.98 m^2 ha^{-1}) and Moishing (332.42 m^2 ha^{-1}) sites (Table 1.3). The density of shrub layer and ground vegetation was highest in the Dambla site (59,000 m^2 ha^{-1}) followed by Changfu Moon (31,800 m^2 ha^{-1}) and Moishing (4,115 m^2 ha^{-1}). The dominance of the tree species was determined using calculated IVI values for each species. In the Changfu Moon site, *Pinus roxburghii*, *Quercus griffithii*, *Juglans regia* and *Rhododendron arboreum* were the dominant tree species, while *Erythrina stricta*, *Schima khasiana* and *Juglans regia* were dominant in the Dambla site. The Moishing site was dominated by *Quercus griffithii* and the bamboo species *Phyllostachys mannii*. Data for density and IVI for each of the associated species is presented in Table 1.3.

The Sorenson similarity index value among the three study sites varied greatly (Table 1.4). The highest similarity index of tree species (80%) was found between the Changfu Moon site and the Dambla site. The similarity index value between these two sites in terms of shrubs/herbs was 23.26% (Table 1.4). The Dambla and Moishing sites were the most dissimilar in tree species content (55.56%) while the herb/shrub richness among the three sites was similar in all the three (Table 1.4).

The density of *G. assamicus* was very low in all sites and ranged between 0.5 (Moishing) to 1 (Changfu Moon and Dambla) individuals per ha (Table 1.5). The IVI of the tree was higher in the Changfu Moon site than in the other two sites (Table 1.5). Seedlings, saplings and tree population varied greatly within the three

Table 1.4. Similarity (Sorenson) among the tree, shrub/herb species occurring in the two study sites.

		Changfu Moon		Dambla Village	
Site		T	S/H	T	S/H
Moishing	T	63.16	–	55.56	–
	S/H	–	22.22	–	23.26
Changfu Moon	T	–	–	80.00	–
	S/H	–	–	–	23.26

Table 1.5. Population status of *G. assamicus* trees in the three study sites.

Site	Density ha^{-1}	IVI
Changfu Moon	1	17.31
Dambla village	1	9.05
Moishing village	0.5	7.70

IVI = Importance Value Index.

Table 1.6. Population status of three *G. assamicus* populations determined from all individuals located during the study.

Parameter	Changfu Moon	Dambla village	Moishing village
Number of individuals	4	8	173
% mature tree	100	87.50	1.16
% saplings (10≥20 cm cd at base and >30 cm height	–	12.50	6.36
% seedlings (≤10 cm cd at base and <30 cm height)	–	–	92.49
Seedling survival (%)	–	–	80–83
New recruitment (%)	–	–	11–14

cd = collar diameter.

sites. In the Changfu Moon site, 100% *G. assamicus* were mature trees while only 1.16% of trees were mature in the Moishing site (Table 1.6). The proportion of saplings at the Dambla and Moishing sites was found to be 12.50% and 6.36%. However, seedling population was recorded only in the Moishing site and constituted 92.49% of the total *G. assamicus* population. Results of the demographic survey of three *G. assamicus* populations are summarized in Tables 1.5 and 1.6.

Present observation revealed that the percentage of mature tree populations was higher than that of sapling/seedling populations in the Dambla and Changfu Moon sites. The largest seedling (160) and sapling (11) populations were recorded in the Moishing site while only one sapling was found in the Dambla site (Fig. 1.14). In the Moishing site, 55% of seedlings occurred within 4 to 8 m radial distance from the mother plant. The density and survival of saplings was recorded only at 8 to 12 m radial distance from the base of the mother plant. There was no seedling or sapling population beyond 12 m radius, suggesting that seed dispersal of *G. assamicus* is very limited.

Our study revealed very few *G. assamicus* individuals in all the studied sites. Though mature trees were present in all sites, regenerating individuals were much less common, and sometimes absent (as in the Changfu Moon and Dambla villages). Rogers (1996) reported similar results in the case of the threatened tree *Olearia hectorii* in New Zealand. Subash Chandran *et al.* (2008) found small and relic populations of two critically endangered tree species *Madhuca bourdillonii* (Gamble) Lam.

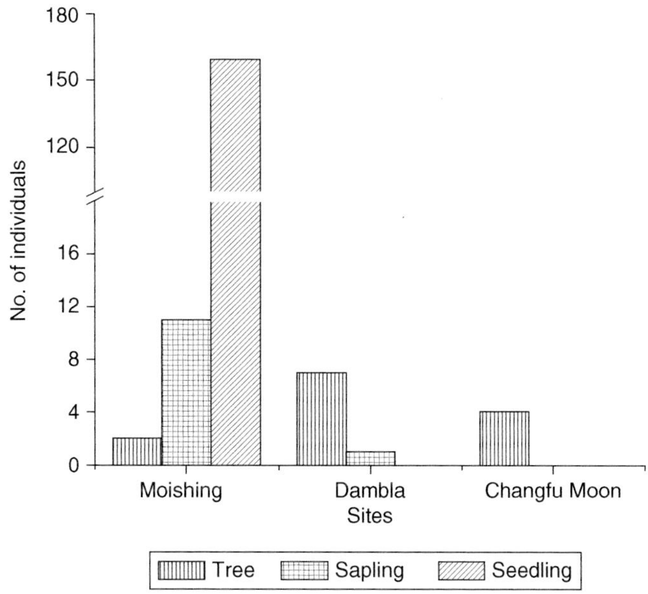

Fig. 1.14. Total number of trees, saplings and seedlings of *G. assamicus* that were recorded in the Moishing, Dambla and Changfu Moon sites.

and *Syzygium travancoricum* Gamble, in Western Ghats, India. The regeneration status of a plant species can be predicted by determining the population size of seedlings and saplings (Khan *et al.*, 1987; Bhuyan *et al.*, 2003). A species may be designated as 'regenerating good' if seedlings > saplings > adults; or fair regenerating if seedlings > or ≤ saplings ≤ adults; and poor regenerating if the species survives only in sapling stage, but no seedlings (saplings may be <, > or = adults – Khan *et al.*, 1987). If a species is present only in adult form, it is considered as 'not regenerating' (Devi, 2004). The age structure of *G. assamicus* indicates that it has very poor regeneration in the Changfu Moon and Dambla sites. The largest seedling and sapling population was recorded in the Moishing site (160 and 11 respectively) and thus regeneration was found to be 'good'. Similar discontinuous population structure has also been reported for a number of tree species such as *Endospermum medullosum* in Solomon Island (Whitmore, 1984), *Trema micrantha* in Barro Colorado Island, Panama (Brokaw, 1987), and *Acer opalus* subsp. *granatense* in the Iberian Peninsula (Gomez-Aparicio *et al.*, 2005).

Though 55% of recorded seedlings were found within a 4–8 m radial distance, density and survival of saplings was recorded only at 8–12 m radial distance and is located away from the crown of the mother tree. Seedling population was found to be significantly ($p<0.05$) higher at 4–8 m radius under the parent tree canopy, whereas sapling population was significantly higher ($p<0.05$) at 8–12 m radial distance (Fig. 1.15). This follows Janzen and Connell's 'escape hypothesis' which states that tree seedlings survive better at greater distances from the parent plant (Janzen,

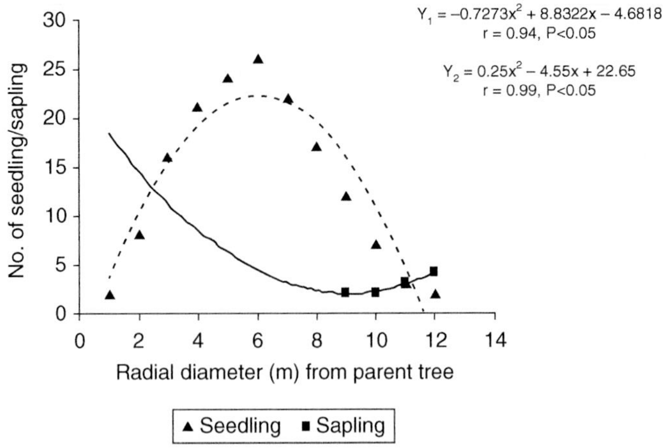

Fig. 1.15. Relationship between radial increase and seedling (Y_1)/sapling (Y_2) survival.

1970; Connell, 1971). Survival at a greater distance may be due to lesser competition among seedlings for growth, and escape from the parent tree. Spatial differences in micro-environment could also be the determining factors behind these differences.

Sapling population size was recorded only from the Dambla and Moishing sites while no sapling was found in the Changfu Moon site. A total of 11 saplings were recorded in the Moishing site while only a single sapling was recorded in the Dambla site. The Changfu Moon site may experience less regeneration due to a lack of fruiting as the site had no hermaphroditic trees, and there is no fruit production in male trees (Table 1.2).

Although there are reports that the survival of tree seedlings is greater in the disturbed forest compared to protected forest (Tripathi and Khan, 1992), present observations revealed that the survival of tree seedlings of *G. assamicus* is greater in less human-impacted forest. This may be due to its remote location, favourable microclimatic conditions and/or its nutrient rich soil. The harvesting of mature fruit, which is a major anthropogenic threat, is also comparatively less in the Moishing site. On the other hand, very poor regeneration in the Dambla site may be due to anthropogenic activities (over-harvesting of mature pods, forest clearance and grazing) due to its proximity to human settlements. However, human disturbances are less prominent at Moishing primarily due to its remote location (Choudhury *et al.*, 2007b).

The survival of seedlings and saplings is greatly impacted by the microclimate, edaphic and biotic factors (Castro *et al.*, 2004). While low temperature is a major abiotic constraint to seedling establishment (James *et al.*, 2004), invertebrate herbivory and pathogenic activity are major biotic constraints (Burdon *et al.*, 1994; Seager *et al.*, 2013). Solar irradiance also impacts seedling establishment (Fisher and Gardner, 1995). Additionally, nutrients play a significant role in seedling growth and survival (Kelly, 2002). Interaction between such biotic and abiotic factors exerts a powerful impact on the spatial pattern of recruitment, concentrating regeneration in some microhabitats and preventing seedling establishment in others (Zackrisson *et al.*, 1997). Moreover, high soil moisture content, higher soil organic C and total Kjeldahl N in the Moishing site may have favoured better regeneration of *G. assamicus.*

During the three years of observation, it was found that 80–83% of seedlings survived in the Moishing site. Additionally, new recruitment of seedlings (11–14%) in the Moishing site was recorded during each study year (Table 1.6). Winter drought due to lack of rainfall coupled with chilly weather between November and March (Fig. 1.16) may be a reason for seedling mortality (Castro *et al.*, 2004). Because of the slow growth rate of *G. assamicus*, no seedling could have been found to attain sapling stage during the three-year study period. Herbivory is an important biotic influence on seedlings because of their size and meal availability to animals (Swaine, 1996). Herbivory was observed as common in all *G. assamicus* sites and may be contributing to survival and growth of the seedling in all studied sites.

The present investigation showed that the *G. assamicus* population is extremely small, and the total number of reproducing individuals is even smaller. Therefore, the species may be designated as 'critically endangered' (Frankel *et al.*, 1995). Very few seedlings, saplings and mature *G. assamicus* trees at the study sites indicates that *G. assamicus* has very poor natural regeneration and survival potential. Other such relict and isolated populations have also experienced decline in recent centuries due to human activity (Blanca *et al.*, 1998; Goettsch *et al.*, 2015). Moreover, no seedling or sapling population has been recorded beyond a 12 m radius from the parent tree, which suggests that seed dispersal of *G. assamicus* is very limited. As dispersal by scatter-hoarding rodents and birds has important implications for the natural regeneration of many tree species (Jansen and Forget, 2001; Niederhauser and Matlack, 2017), the lack of dispersal observed in *G. assamicus* may act as a potential regeneration constraint.

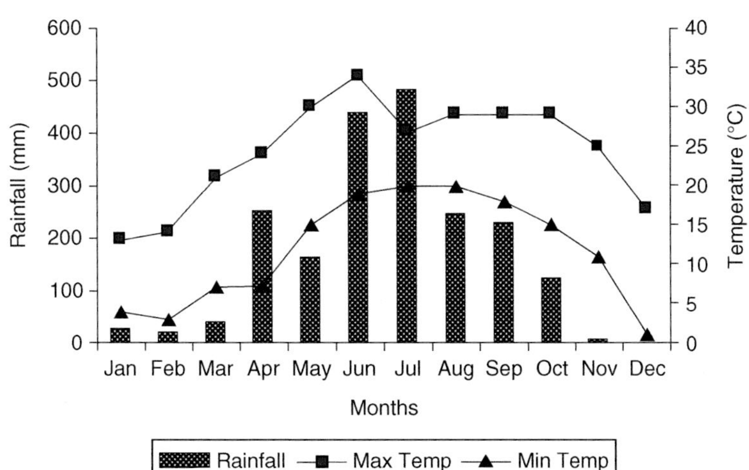

Fig. 1.16. Ombrothermic diagram of the study area. Mean monthly rainfall (mm) and maximum and minimum temperature (°C) for 2004–2006.

1.6 Microclimatic Conditions

The survival and growth of a species largely depends on various ecological conditions and is largely influenced by microenvironmental variables and microsite heterogeneity. Climate dominantly exerts control over the distribution of major vegetation types (Woodward, 1987). Each and every stage of the plant life cycle, such as leaf flushing, flowering, pollination and fertilization, fruit set and maturation, seed dispersal, germination and survival of seedlings, is influenced by the climate. Alteration of climatic conditions between years affects seed production significantly (Owens, 1995). Aussenac (2000) studied the interactions between forest stands and microclimate, and found marked influence on forest trees. For example, the survival and growth of seedlings of a few tree species was found to be higher in gaps as compared with the understory, and the prevailing microenvironmental conditions may have favoured the growth and establishment of the seedlings (Khumbongmayum *et al.*, 2005).

To understand the various microclimatic conditions, we measured photosynthetic active radiation (PAR) in the study sites using plant canopy analyser (LI-COR, Model LI-250, US made) between March and April (2005) at 08.00, 12.00 and 16.00 h for seven days. The mean quantum of light recorded at the ground level varied significantly ($F=108399.635$; $p<0.001$) within the study sites with maximum values in the Moishing site and minimum in the Changfu Moon site. PAR ranged between 242–250, 1,123–1,130 and 38–45 μ mol s^{-1} m^{-2} in Changfu Moon, Dambla village and Moishing village respectively (Table 1.1).

Soil properties of a land under pasture with a 17-year old naturally regenerated secondary forest were studied by Alfaro (2000), who found that soil under forest area had better conductivity. Variation in community structure in a plantation has been attributed to a soil fertility gradient, defined by increasing pH and available soil nutrients, that is, available N in NO_3 (Sharma, 1993). Daws *et al.* (2002) in Barro Colorado Island found that large variations in soil water regime over small spatial scales may play a role in maintaining high species richness by opening up ecological niches.

Physico-chemical properties of soil such as colour, texture, bulk density, water-holding capacity, soil moisture content, pH, organic carbon and total Kjeldahl nitrogen (TKN) were analysed. Triplicate soil samples were collected in April 2005 from each site using a steel corer (5.5 cm inner diameter and 30 cm height) from two soil depths (0–15 and 15–30 cm) after removing the litter layer. The replicated samples of each site and depth were thoroughly mixed to obtain a composite sample. The samples were brought to the laboratory and select subsamples were analysed for soil moisture and pH in fresh conditions. Remaining soil samples were air dried, passed through a 2 mm mesh sieve to remove stone particles and gravels and then again sieved through 0.5 mm mesh screen for determination of chemical parameters. Separate soil samples were collected from each site for determination of soil bulk density.

Soil colour was determined with the help of the Munsell soil colour chart. Soil texture was determined by the Bouyoucos hydrometer method (Bouyoucos, 1962) and bulk density was determined by the soil core method (Blake and Hartge, 1986).

Water-holding capacity (WHC) was determined following the Keen's box method (Piper, 1944), while soil moisture content was measured gravimetrically by incubating 10 g of field-moist soil sample in a hot air oven at 105°C for 24 h. Soil organic carbon (SOC) was determined by dichromate oxidation and titration with ferrous ammonium sulphate (Walkey, 1947). TKN was determined following semi-micro Kjeldahl procedure by acid digestion, distillation and titration (Anderson and Ingram, 1993). The pH of soil samples was determined through the soil–water suspension method (1:2.5 w/v) using a digital pH meter (Systronics MK VI).

All the data were statistically analysed following Zar (1974), using multiway ANOVA to elucidate the effects of site characteristics on distribution, growth and other phytosociological parameters of *G. assamicus*.

All the sites are located within an elevation range of 1,700 and 2,100 m asl and have either a northwest or southeast aspect. The mean PAR varied between 38.34 μ mol s^{-1} m^{-2} in the Moishing site to 1123.65 μ mol s^{-1} m^{-2} in the Dambla site. Lesser quantum of light in the Moishing site may be due to partially closed canopy cover and less disturbance, while more quantum in the Dambla site may be due to fully open canopy cover. Similar variation has also been reported by Chen *et al.* (1999).

The soil in all study sites was classified as sandy loam, having 67.67–70.33% sand with a small proportion of clay (Table 1.7). Soil colour varied among the sites. In the Dambla site, soil colour was brown and dark brown, whereas in the Changfu Moon and Moishing sites, it was dark grey to very dark grey (Table 1.7). Soil moisture content ranged from 17.10–30.71% with a maximum value in the Moishing site, and a minimum value in the Dambla site. Variation in soil moisture content was significant between sites (F= 3571.710; $p<0.001$), and insignificant between the two soil depths. WHC ranged between 46.07–67.32%; the

Table 1.7. Soil physical properties of the three study sites.

Site	Changfu Moon		Dambla village		Moishing	
Soil depth (cm)	0–15	15–30	0–15	15–30	0–15	15–30
Texture						
Sand (%)	67.67 ±1.76	70.33 ±1.76	67.67 ±1.76	69.00 ±2.00	67.00 ±1.15	70.33 ±1.76
Silt (%)	24.00 ±2.08	23.33 ±2.03	26.00 ±1.73	25.67 ±2.33	23.67 ±1.45	23.00 ±1.73
Clay (%)	8.33 ±0.33	6.33 ±0.33	6.33 ±0.33	5.33 ±0.33	9.33 ±0.33	6.67 ±0.33
Textural class	SL	SL	SL	SL	SL	SL
WHC (%)	67.32 ±0.07	65.37 ±0.13	48.43 ±0.32	46.07 ±0.47	61.37 ±0.16	60.61 ±0.11
BD (g cm^{-3})	0.83 ±0.003	0.87 ±0.002	0.98 ±0.007	1.01 ±0.005	0.67 ±0.006	0.72 ±0.003
SMC (%)	22.45 ±0.19	23.51 ±0.11	17.38 ±0.21	17.10 ±0.05	31.67 ±0.24	30.71 ±0.12

± SE (n=3); WHC: water-holding capacity; BD: bulk density; SMC: soil moisture content; SL = sandy loam.

highest value was recorded in the the Changfu Moon site (67.32%) and the lowest (46.07%) in the Dambla site (Table 1.7). Variation in moisture and WHC of soil is due to the change in soil texture within different sites.

Soil was medium acidic to strongly acidic (pH 5.97–4.71) in all the three sites, with maximum acidity (pH=4.71) being recorded in the Moishing site. The pH value varied significantly between sites (F=230.311, p<0.001), whereas the difference was insignificant between soil depths. Soil pH was generally lower in the surface soil layer than in the subsurface layer. Low soil pH in the Moishing site as compared with the Changfu Moon and Dambla sites could be due to a lower rate of soil leaching, leading to greater accumulation of partially decomposed organic matter. An increase in pH in the other sites was due to low organic matter accumulation. The SOC varied significantly among the sites, as well as between the two soil depths (F= 246.112, 186.483; p<0.001). Similarly, total Kjeldahl nitrogen varied significantly between the two soil depths and sites (F=627.706, 1409.086; p<0.001). The highest concentration of TKN was recorded in the surface soil layer of the Moishing site, while the lowest concentration was recorded in the subsurface soil layer of the Dambla site (Table 3.3). Higher soil organic C and total Kjeldahl N in the Moishing site may be due to lesser disturbance as compared with the Changfu Moon and Dambla sites.

Disturbance in forest areas often changes the amount and quality of litter and hence reduces organic matter input and nutrient content of the soil (Dinesh *et al.*, 2003). Soil texture is also greatly affected by the openness of the canopy due to the erosion of topsoil during extreme rainfall events, and increases the proportion of sand in the disturbed stands due to loss of finer soil particles (Eyre, 1968). Loss of canopy trees and extreme rainfall events might have increased the erosion of topsoil layer and hence increased the proportion of sand particles in the Dambla and Changfu Moon sites. The Moishing site was found to have the highest clay and moisture content compared to the other sites, which may be due to lesser disturbance, low solar insolation and the presence of a seasonal stream. Highest moisture content in the surface soil layer may be due to greater organic content in this layer. WHC in all the sites was greater in the surface soil layer, which may be due to relatively higher clay content and the accumulation of organic matter in the topsoil layer. Higher bulk density in the Dambla site may be due to loss of organic matter, as the site experiences significantly higher levels of disturbance as compared with the other sites.

1.7 References

Alfaro, R.E., 2000. Improvement in the physical and chemical properties of soils during the process of natural regeneration at the Estacion Experimental Forestal Horizontes (EEFH), Guanacaste, Costa Rica. In *Taller de Nutricion Forestall, Memoria, 2 de junio*. San Jose, Costa Rica. Consejo Nacionalde Rectores, pp. 76–95.

Anderson, J.M. and Ingram, J.S.I., 1993. *Tropical Soil Biology and Fertility: A Handbook of Methods*. CAB International, Wallingford, UK.

Armesto, J.J. and Pickett, S.T.A., 1985. Experiments on disturbance in old-field plant communities: impact on species richness and abundance. *Ecology* 66(1): 230–240.

Aussenac G., 2000. Interactions between forest stands and microclimate: ecophysiological aspects and consequences for silviculture. *Annals of Forest Science* 57: 287–301.

Batalden, R.V., Oberhauser, K. and Peterson, A.T., 2014. Ecological niches in sequential generations of eastern North American monarch butterflies (Lepidoptera: Danaidae): the ecology of migration and likely climate change implications. *Environmental Entomology* 36(6):1365–1373.

Bawri, A., Gajurel, P.R. and Khan, M.L., 2016. *Primula calderiana* subsp. *bawaii* (Primulaceae): a new subspecies from Arunachal Pradesh, India. *Phytotaxa*, 265(1): 67–72.

Benthum, G. and Hooker, J.D., 1862–1883. *Genera Plantarum*, 1–3. Lovell Reeve, London, UK.

Bhuyan, P., Khan, M.L. and Tripathi, R.S., 2003. Tree diversity and population structure in undisturbed and human-impacted stands of tropical wet evergreen forest in Arunachal Pradesh, Eastern Himalayas, India. *Biodiversity and Conservation* 12: 1753–1773.

Blake, G.R. and Hartge, K.H., 1986. Bulk density. In A. Klute (ed.) *Methods of Soil Analysis, Part I. Physical and Mineralogical Methods.* American society of Agronomy Monograph no. 9 (2nd edn), pp. 363–375.

Blanca, G., Cueto, M., Martinez-Lirola, M.J. and Molero-Mesa, J., 1998. Threatened vascular flora of Sierra Nevada (Southern Spain). *Biological Conservation* 85: 269–285.

Bourg, N.A., McShea, W.J. and Gill, D.E., 2005. Putting a CART before the search: successful habitat prediction for a rare forest herb. *Ecology* 86(10): 2793–2804.

Bouyoucos, G.J., 1962. Hydrometer method improved for making particle size analyses of soils. *Agronomy Journal* 54: 464–465.

Bowles, J.M., 2004. *Special Plant Abstract for Gymnocladus dioicus (Kentucky Coffee-tree).* Walpole Island Heritage Centre and Michigan Natural Features Inventory, Lansing, MI, p. 3.

Brokaw, N.V.L., 1982. Treefalls: frequency, timing and consequences. In A.S. Leigh, E.G. Rand and D.M. Windsor (eds) *The Ecology of a Tropical Forest.* Smithsonian Institution Press, Washington, DC, pp.101–108.

Brokaw, N.V.L., 1987. Gap phase regeneration of three pioneer tree species in a tropical forest. *Journal of Ecology* 75: 9–19.

Burdon, J.J., Wennstrom, A., Muller, W.J. and Ericson, L., 1994. Spatial pattering in young stands of *Pinus sylvestris* in relation to mortality caused by the snow blight pathogens *Phacidium infestans*. *Oikos* 71: 130–136.

Castro, J., Zamora, R., Hodar, J.A. and Gomez, J.M., 2004. Seedling establishment of a boreal tree species (*Pinus sylvestris*) at its southernmost distribution limit: consequences of being in a marginal Mediterranean habitat. *Journal of Ecology* 92: 266–277.

Champion, S.H. and Seth, S.K., 1968. *A revised survey of the forest types of India.* Manager of Publications, Delhi (e-book).

Chen, J., Sari, C.S., Crow, T.R., Naiman, R.J., Brosofske, K.D., Mroz, G.D., Brookshire, B.L. and Franklin, J.F., 1999. Microclimate in forest ecosystem and landscape ecology. *Bioscience* 49: 288–297.

Choudhury, B.I., Khan, M.L., Arunachalam, A. and Arunachalam, K., 2007a. *Gymnocladus assamicus* Kanjilal ex PC Kanjilal fruit – a soap substitute. *Natural Product Radiance* 6: 427–429.

Choudhury, B.I., Khan, M.L., Arunachalam, A. and Das, A.K., 2007b. Ecology and conservation of the critically endangered tree species *Gymnocladus assamicus* in Arunachal Pradesh, India. *International Journal of Ecology* 2007.

Chowdhery, H.J. and Pal, G.D., 1997. A new species of Coelogyne (Orchidaceae) from Arunachal Pradesh, India. *Nordic Journal of Botany* 17(4): 369–371.

Clements, F.E., 1936. Nature and structure of the climax. *Journal of Ecology* 24: 252–284.

Connell, J.H., 1971. On the role of natural enemies in preventing competitive exclusion in some marine animals and in rain forest trees. In P.J. Den Boer and G. Gradwell G. (eds) *Dynamics of Populations*. PUDOC, Wageningen.

Corlett, R.T. and Westcott, D.A., 2013. Will plant movements keep up with climate change? *Trends in Ecology & Evolution* 28(8): 482–488.

Cox, C.B. and Moore, P.D. 1993. *Biogeography: An Ecological and Evolutionary Approach* (5th edn). Blackwell Scientific, London.

Daws, M.I., Mullins, C.E., Burslem, D.F.R.P., Paton, S.R. and Dalling, J.W., 2002. Topographic position affects the water regime in a semideciduous tropical forest in Panama. *Plant and Soil* 238: 79–90.

Devi, K.A., 2004. Studies on plant diversity and regeneration of a few tree species in the Sacred Groves of Manipur. PhD thesis, North-Eastern Hill University, Shillong, India.

Dinesh, R., Ghosal, C.S., Ganeshamurthy, A.N. and Dey, C., 2003. Changes in soil microbial indices and their relationships following deforestation and cultivation in wet tropical forests. *Applied Soil Ecology* 24: 17–26.

Dong, S., Chettri, N. and Sharma, E., 2017. Himalayan biodiversity: trans-boundary conservation, institution and governance. In S. Dong, J. Bandyopadhyay and S. Chaturvedi (eds) *Environmental Sustainability from the Himalayas to the Oceans* . Springer International Publishing, New York, pp. 127–143.

Eyre, S.R., 1968. *Vegetation and Soil: A World Picture*. Edwards Arnold Publishers Ltd, UK.

Fisher, M. and Gardner, A.S., 1995. The status and ecology of *Juniperus excelsa* subsp. *polycarpos* woodland in the northern mountains of Oman. *Vegetation* 119: 33–51.

Franco, J.A., 1986. *Taxus baccata* L. In S. Castroviejo, M. Lanz, G. Lopez, P. Monserrat, F. Munoz, J. Paiva and L. Villar (eds) *Flora Iberica, Vol. I: Lycopodiaceae-Pappaveraceae*. Real jardin Botanico, Servicio de Publications del C.S.I.C., Madrid, pp. 201–202.

Frankel, O.T., Brown, A.H.D. and Burdon, J.J., 1995. *The Conservation of Plant Diversity*. Cambridge University Press, Cambridge, p. 299.

Ganeshaiah, K.N., 2005. Recovery of endangered and threatened species: developing a national priority list of plants and insects. *Current Science* 89(4): pp. 599–600.

Garcia, D., Zamora, R., Hodar, J.A., Gomez, J.M. and Castro, J., 2000. Yew (*Taxus baccata* L.) regeneration is facilitated by fleshy-fruited shrubs in Mediterranean environments. *Biological Conservation* 95: 31–38.

Goettsch, B., Hilton-Taylor, C., Cruz-Piñón, G., Duffy, J.P., Frances, A., Hernández, H.M., Inger, R., Pollock, C., Schipper, J., Superina, M. and Taylor, N.P., 2015. High proportion of cactus species threatened with extinction. *Nature plants 1*: 15142.

Gomez-Aparicio, L., Zamora, R. and Gomez, J.M., 2005. The regeneration status of the endangered *Acer opalus* subsp. *granatense* throughout its geographical distribution in the Iberian Peninsula. *Biological Conservation* 121: 195–206.

Good, N.F. and Good, R.E., 1972. Population dynamics of tree seedlings and saplings in mature eastern hardwood forests. *Bulletin of Torrey Botanical Club* 99: 172–178.

Götmark, F. and Kiffer, C., 2014. Regeneration of oaks (*Quercus robur/ Q. petraea*) and three other tree species during long-term succession after catastrophic disturbance (windthrow). *Plant ecology* 215(9): 1067–1080.

Guisan, A. and Zimmermann, N.E., 2000. Predictive habitat distribution models in ecology. *Ecological modelling* 135(2): 147–186.

Hajra, P.K., Verma, D.M. and Giri, G.S., 1996. *Materials for the flora of Arunachal Pradesh*. Botanical Survey of India, Calcutta, 693 pp.

Hegde, S.N., 2002. *Arunachal Pradesh State Biodiversity Action Plan.* Arunachal Pradesh Forest Research and Development Agency, State Forest Research Institute, Itanagar, Arunachal Pradesh, India.

Hirayama, K. and Sakimoto, M., 2003. Regeneration of *Cryptomeria japonica* on a sloping topography in a cool-temperate mixed forest in the snowy region of Japan. *Canadian Journal of Forest Research* 33: 543–551.

Holeksa, J., Jaloviar, P., Kucbel, S., Saniga, M., Svoboda, M., Szewczyk, J., Szwagrzyk, J., Zielonka, T. and Żywiec, M., 2017. Models of disturbance driven dynamics in the West Carpathian spruce forests. *Forest Ecology and Management* 388: 79–89.

James, J.C., Grace, J. and Hoad, S.P., 2004. Growth and photosynthesis of *Pinus sylvestris* at its altitudinal limit in Scotland. *Journal of Ecology* 82: 297–306.

Jansen, P.A. and Forget, P.M., 2001. Scatter-hoarding rodents and tree regeneration. In F. Bongers, P. Charles-Dominique, P.M. Forget and M. Théry (eds) *Nouragues: Dynamics and Plant-Animal Interactions in a Neotropical Rainforest.* Kluwer Academic Publishers, Dordrecht, the Netherlands, pp. 275–288.

Janzen, D.H., 1970. Herbivores and the number of tree species in tropical rain forests. *American Naturalist* 104: 501–528.

Jathar, G., Patil, D., Kalra, M., de Silva, T., Peterson, A.T., Irfan-Ullah, M., Rahmani, A.R., Mehta, P. and Kulkarni, J., 2015. Mapping the potential distribution of the critically endangered forest owlet *Heteroglaux blewitti* in India. *Journal of the Bombay Natural History Society (JBNHS)* 112(2): 55–64.

Justice, C.O., Vermote, E., Townshend, J.R., Defries, R., Roy, D.P., Hall, D.K., Salomonson, V.V., Privette, J.L., Riggs, G., Strahler, A. and Lucht, W., 1998. The Moderate Resolution Imaging Spectroradiometer (MODIS): land remote sensing for global change research. *IEEE Transactions on Geoscience and Remote Sensing* 36(4): 1228–1249.

Kanjilal, U.N., Kanjilal, P.C. and Das, A., 1938. *Flora of Assam.* Vol. I–IV. Government of Assam, Shillong.

Kaul, R.N. and Haridasan, K., 1987. Forest types of Arunachal Pradesh: a preliminary study. *Journal of Economic and Taxonomic Botany* 9(2): 379–389.

Kelly, D.L., 2002. The regeneration of *Quercus petrea* (sessile oak) in southwest Ireland: a 20-year experimental study. *Forest Ecology and Management* 166: 207–226.

Khan, M.L. and Tripathi, R.S., 1986. Tree regeneration in a disturbed sub-tropical wet hill forest of north-east India: effect of stump diameter and height on sprouting of four tree species. *Forest Ecology and Management* 17(2–3): 199–209.

Khan, M.L., Rai, J.P.N. and Tripathi, R.S., 1987. Population structure of some tree species in disturbed and protected subtropical forests of northeast India. *Acta Oecologica* 8(3): 247–255.

Khumbongmayum, A.D., Khan, M.L. and Tripathi, R.S., 2005. Biodiversity conservation in sacred groves of Manipur, northeast India: population structure and regeneration status of woody species. *Biodiversity and Conservation* 14(7): 1541–1582.

Kohyama, T. and Hara, T., 1989. Frequency distribution of tree growth rate in natural forest stands. *Annals of Botany* 64(1): 47–57.

Konoshima, T., Yasuda, I., Kashiwada, Y., Cosentino, L.M. and Lee, K.H., 1995. Anti-AIDS agents, 21. triterpenoid saponins as Anti-HIV principles from fruits of *Gleditsia japonica* and *Gymnocladus chinesis*, and a structure-activity correlation. *Journal of Natural Products*, 58(9): 1372–1377.

Lee, K.H. and Morris-Natschke, S.L., 1999. Recent advances in the discovery and development of plant-derived natural products and their analogs as anti-HIV agents. *Pure and Applied Chemistry*, 71(6): 1045–1051.

Mehta, P.S. and Bhatt, K.C., 2007. Traditional soap and detergent yielding plants of Uttaranchal. *Indian Journal of Traditional Knowledge* 6(2): 279–284.

Menon, S., Choudhury, B.I., Khan, M.L. and Peterson, A.T., 2010. Ecological niche modeling and local knowledge predict new populations of *Gymnocladus assamicus* a critically endangered tree species. *Endangered Species Research* 11(2): 175–181.

Menon, S., Latif Khan, M., Paul, A. and Peterson, A.T., 2012. Rhododendron species in the Indian Eastern Himalayas: new approaches to understanding rare plant species distributions. *Journal American Rhododendron Society* 66(2): 78–84.

Miller, A.D., Roxburgh, S.H. and Shea, K., 2011. How frequency and intensity shape diversity–disturbance relationships. *Proceedings of the National Academy of Sciences* 108(14): 5643–5648.

Misra, R., 1968. *Ecology Work Book*. Oxford and IBH Publishing Co., New Delhi.

Nakayama, K., Fujino, H., Kasai, R., Mitoma, Y., Yata, N. and Tanaka, O., 1986. Solubilizing properties of saponins from *Sapindus mukurossi* Gaertn. *Chemical and Pharmaceutical Bulletin*, 34(8): 3279–3283.

Nath, P.C., Arunachalam, A., Khan, M.L., Arunachalam, K. and Barbhuiya, A.R., 2005. Vegetation analysis and tree population structure of tropical wet evergreen forests in and around Namdapha National Park, northeast India. *Biodiversity and Conservation* 14(9): 2109–2136.

Niederhauser, E.C. and Matlack, G.R., 2017. Do deer and raccoons defecate in the right place? Fitness consequences of vertebrate seed dispersal for a deciduous forest herb. *Oecologia* 183(3): 727–737.

Osbourn, A., 1996. Saponins and plant defence – a soap story. *Trends in Plant Science* 1(1): 4–9.

Owens, J.N., 1995. Constraints to seed production; temperate and tropical forest trees. *Tree Physiology* 15: 747–404.

Peterson, T.A., Papeş, M. and Eaton, M., 2007. Transferability and model evaluation in ecological niche modeling: a comparison of GARP and Maxent. *Ecography* 30(4): 550–560.

Phillips, S.J., Anderson, R.P. and Schapire, R.E., 2006. Maximum entropy modeling of species geographic distributions. *Ecological modelling*, 190(3): 231–259.

Piper, C.S., 1944. *Soil and Plant Analysis*. John Wiley & Sons, New York, USA.

Polunin, O. and Stainton, A., 2006. *Flowers of the Himalaya*. Oxford University Press, 8th Impression, Oxford.

Raghavan, R.K., Goodin, D.G., Hanzlicek, G.A., Zolnerowich, G., Dryden, M.W., Anderson, G.A. and Ganta, R.R., 2016. Maximum entropy-based ecological niche model and bio-climatic determinants of Lone Star Tick (*Amblyomma americanum*) niche. *Vector-Borne and Zoonotic Diseases* 16(3): 205–211.

Rao, P.B., 1988. Effects of environmental factors on germination and seedling growth in *Quercus floribunda* and *Cupressus torulosa* tree species in central Himalaya. *Annals of Botany* 61: 531–540.

Rappole, J.H., Renner, S.C., Shwe, N.M. and Sweet, P.R., 2005. A new species of scimitar-babbler (Timaliidae: Jabouilleia) from the sub-Himalayan region of Myanmar. *The Auk* 122(4): 1064–1069.

Raxworthy, C.J., Martinez-Meyer, E., Horning, N., Nussbaum, R.A., Schneider, G.E., Ortega-Huerta, M.A. and Peterson, A.T., 2003. Predicting distributions of known and unknown reptile species in Madagascar. *Nature* 426(6968): 837–841.

Roff, D. (ed.), 1993. *Evolution of Life Histories: Theory and Analysis*. Springer Science & Business Media, New York.

Rogers, G.M., 1996. Aspects of the ecology and conservation of the threatened tree *Olearia hectorii* in New Zealand. *New Zealand Journal of Botany* 34: 227–240.
Saha, D. and Sundriyal, R.C., 2012. Utilization of non-timber forest products in humid tropics: implications for management and livelihood. *Forest Policy and Economics* 14(1): 28–40.
Saha, D., Ved, D., Ravikumar, K. and Haridasan, K., 2015. *Gymnocladus assamicus*. The IUCN Red List of Threatened Species 2015, e.T50126611A50131355. http://dx.doi.org/10.2305/IUCN.UK.2015-2.RLTS.T50126611A50131355.en, accessed 11 June 2017.
Sanjappa, M., 2000. Checklist of the Leguminosae of South Asia. *ILDIS World Database of Legumes* Ver. 6, 4th Intl. Legume Conf., Canberra, Australia.
Seager, S.T., Eisenberg, C. and Clair, S.B.S., 2013. Patterns and consequences of ungulate herbivory on aspen in western North America. *Forest Ecology and Management*, 299: 81–90.
Segura, G., Balvanera, P., Duran, E. and Perez, A., 2003. Tree community structure and stem mortality along a water availability gradient in a Mexican tropical dry forest. *Plant Ecology* 169: 259–271.
Sharma, E., 1993. Nutrient dynamics in Himalayan alder plantations. *Annals of Botany* 72: 329–336.
Soberón, J. and Peterson, A.T., 2005. Interpretation of models of fundamental ecological niches and species' distributional areas. Biodiversity Informatics, https://doi.org/10.17161/bi.v2i0.4, accessed 2 April 2019.
Sorenson, T., 1948. A method of establishing groups of equal amplitude on similarity of species content. *Biologiske Skrifter K. Danske Videnskbernes Selskab* 5(4): 1–34.
Stainton, A., 2007. *Flowers of the Himalaya. A Supplement.* Oxford University Press, Oxford.
Stockwell, D., 1999. The GARP modelling system: problems and solutions to automated spatial prediction. *International Journal of Geographical Information Science* 13(2): 143–158.
Subash Chandran, M.D., Mesta, D.K., Rao, G.R., Ali, S., Gururaja, K.V. and Ramachandra, T.V., 2008. Discovery of two critically endangered tree species and issues related to relic forests of the Western Ghats. *The Open Conservation Biology Journal* 2: 1–8.
Suhagia, B.N., Rathod, I.S. and Sindhu, S., 2011. *Sapindus mukorossi* (Areetha): an overview. *International Journal of Pharmaceutical Sciences and Research* 2(8): 1905.
Swaine, M.D., 1996. Forword. In M.D. Swaine (ed.) *The Ecology of Tropical Forest Tree Seedlings.* Man and Biosphere Series. UNESCO/Parthenon, Paris/Carnforth, 17.
Swaine, M.D. and Hall, J.B., 1983. Early succession on cleared forest land in Ghana. *Journal of Ecology* 71: 601–627.
Tamuly, C., Hazarika, M. and Bordoloi, M., 2013. Biosynthesis of Au nanoparticles by *Gymnocladus assamicus* and its catalytic activity. *Materials Letters* 108: 276–279.
Thom, D. and Seidl, R., 2016. Natural disturbance impacts on ecosystem services and biodiversity in temperate and boreal forests. *Biological Reviews* 91(3): 760–781.
Tingstad, L., Olsen, S.L., Klanderud, K., Vandvik, V. and Ohlson, M., 2015. Temperature, precipitation and biotic interactions as determinants of tree seedling recruitment across the tree line ecotone. *Oecologia* 179(2): 599–608.
Tripathi, R.S. and Khan, M.L., 1992. Regeneration pattern and population structure of trees in subtropical forest of northeast India. In K.P. Singh and J.S. Singh (eds) *Tropical Ecosystems: Ecology and Management.* Wiley Eastern Limited, New Delhi, pp. 43–441.
USDA, 2017. Kentucky Coffeetree (*Gymnocladus dioicus*) (PDF). https://plants.usda.gov/core/profile?symbol=GYDI. Retrieved 11 June 2017.

Valdiya, K.S., 2002. Geodynamic perspective of Arunachal Pradesh: bearing on environmental security and planning for development. *HIMAVIKAS Occasional Publication*, (16): 1–15.

Venugopal, N. and Pamidimarri, S., 2015. Rediscovery of critically endangered *Gymnocladus assamicus* Kanjilal *ex* PC Kanjilal (Leguminosae: Caesalpinioideae) from West Khasi Hills District of Meghalaya. *Pleione* 9(1): 186–192.

Walkey, A., 1947. A critical examination of a rapid method for determining organic carbon in soils: effect of variations in digestion conditions and inorganic soil constituents. *Soil Science* 63: 251–263.

Watkinson, A.R., 1997. Plant population dynamics. In M.J. Crawly (ed.) *Plant Ecology* (2nd edn). Blackwell, Oxford, pp. 359–400.

Whitmore, T.C., 1984. *Tropical Forest of the Far East.* Claredon Press, Oxford.

Woodward, F.I., 1987. *Climate and Plant Distribution.* Cambridge University Press, London.

Wright, H.E.J. and Heinselman, M.L. 1973. Ecological role of fire. *Quaternary Research* 3: 319–328.

Wulff, A.S., Hollingsworth, P.M., Ahrends, A., Jaffré, T., Veillon, J.M., L'Huillier, L. and Fogliani, B., 2013. Conservation priorities in a biodiversity hotspot: analysis of narrow endemic plant species in New Caledonia. *PLoS One* 8(9): p.e73371.

Zackrisson, O., Nilsson, M.C., Dahlberg, A. and Jaderlund, A., 1997. Interference mechanisms in conifer-Ericaceae-feathermoss communities. *Oikos* 78: 209–220.

Zar, J.H., 1974. *Biostatistical Analysis.* Englewood Cliffs, NJ: Prentice Hall.

2 Phylogeny and Phylogeography of the Genus *Gymnocladus* and its Close Relatives

2.1 Introduction

The genus *Gymnocladus* belongs to the family Leguminosae (alternative name *Fabaceae*) and subfamily Caesalpinioideae, one of the plant families distributed throughout the world that grow in a wide range of habitats. Leguminosae is the third largest angiosperm family consisting of almost 770 genera and over 19,500 species (Doyle and Luckow, 2003; Lewis *et al.*, 2005). This is an incredibly diverse group of plants in almost every form and way one can imagine. For example, wide variation exists in floral morphology, such as wild rose-like appearance in yearly-diverging lineages, the reduced flowers of the Amorpha to the spectacular orchid-like flowers in others (Doyle and Luckow, 2003). Great variation in chromosome numbers is also observed in the Leguminosae and among the subfamilies, tribes and even at the generic level (Bandel, 1974). Members of Leguminosae are almost cosmopolitan in distribution and show enormous diversity in their growth habits from herbaceous to aquatic, from shrubs, vines and lianas to trees, and in physiology and life cycles. Schrire *et al.* (2005a) grouped the plants into four major biomes, namely succulent, grass, rainforest and temperate.

Though there were ambiguities regarding the origin and evolution of legumes and many have predicted that legumes evolved in the humid tropics, fossil records as well as phylogenetic studies using molecular data suggest that legume species originally evolved in arid and semiarid regions along the Tethys seaway and diversified during the Early Tertiary period (Herendeen *et al.*, 1992; Lavin *et al.*, 2005; Schrire *et al.*, 2005b). Legumes represent several primitive traits which are correlated with lower-level plant groups such as ferns as well as angiosperms (Herendeen *et al.*, 1992; Lavin *et al.*, 2005). It is estimated that legumes first appeared during the Late

Paleocene, approximately 56 million years ago (mya) (Wing *et al.*, 2004) and evolved into different clades and subclades over millions of years. Over time, the group has established and rapidly diversified across the globe to become an integral part of the biota. Earlier hypotheses suggested that the family originated in West Gondwana (austrotropical) with subsequent vicariance events involving South American and African lineages (Raven and Axelrod, 1974; Morley, 2000), however, the Tethys Seaway hypothesis is currently the most commonly accepted explanation for legume origin and evolution (Ramdhani *et al.*, 2010; Yahara *et al.*, 2013). Other views also suggest a possible northern Laurasian origin of legume (Bouchenak-Khelladi *et al.*, 2010). Regarding migration, it was thought that legumes directly spread between South America and Africa during continental separation, however, this idea has changed recently to a northern Laurasian origin of legume. Spread of legumes northwards from the Americas via the boreotropics and then southwards to reach Africa and Asia (boreal hypothesis) is also well recognized (Doyle and Luckow, 2003).

Lavin *et al.* (2005) estimated the ages of different clades based on molecular data and showed that the most recent diversification took place about 5.1 mya, thereby highlighting a continuous evolutionary process. Based on the phylogenetic reconstruction and legume phylogenetic tree, Schrire *et al.* (2005b) suggested that the semiarid succulent biome is the parental lineage that gave rise to other sublineages that occupy all other biomes. The current pattern and worldwide distribution of legume indicates more relatedness to biogeographical factors such as rainfall and temperature than to rafting of land masses (Sprent, 2007).

Legume is also considered the second most economically important plant family after graminiae in the plant kingdom. These plants serve as a dominant source of dietary proteins and oils for human consumption (Graham and Vance, 2003). At the same time, legume species diversity has been rapidly depleting due to current threats to biodiversity by natural and anthropogenic agents. As a result, project such as Global Legume Diversity Assessment (GLDA) has been initiated to understand the various challenges and preserve genetic resources for sustainable utilization in the future (Yahara *et al.*, 2013).

Phylogeny of legume is a very complicated subject and remains poorly understood despite several attempts using traditional and molecular approaches. The subfamily Caesalpinioideae is generally classified as paraphyletic (Lewis *et al.*, 2005) and is currently divided into four tribes, two of which (Cercideae and Detarieae) are monophyletic while Cassieae and Caesalpinieae are paraphyletic with Caesalpinieae as a typically difficult tribe from a phylogenetic perspective (Manzanilla and Bruneau, 2012). The family has a basal position in phylogenetic schemes of the angiosperms (Doyle *et al.*, 2000; Bruneau *et al.*, 2001) and is highly diverse in floral form and ontogeny with a complex evolutionary history. Though significant progress has been made in understanding the phylogeny and evolution of the legume family in the last decade, it is often challenging to ascertain robust phylogenetic relationships among different lineages. A more elaborated study of the tribe Caesalpinioideae revealed six informal clades such as the Umtiza, Caesalpinia, Tachigali, Peltophorum and two Dimorphandra group lineages as well as the Cassia clade (Bruneau *et al.*, 2008). Manzanilla and Bruneau (2012) performed a robust phylogenetic study of the Caesalpinieae clades using duplicated copies of the sucrose synthase gene and plastid DNA markers and found four well-supported clades: Cassia, Caesalpinia, Tachigali and Peltophorum. Furthermore, they also showed two paraphyletic groups (Dimorphandra) and Umtiza clade (Manzanilla and Bruneau, 2012) to which *Gymnocladus* species belong.

2.2 Legume Diversity in the Eastern Himalayan Region

Caesalpinioideae has been considered primarily as a tropical subfamily with the majority of the species endemic to the tropical regions of Africa, America and Asia (Lee, 1976). The subfamily Caesalpinioideae encompasses 170 genera and about 3,000 species (Tucker, 2003). The Amazon basin has been suggested as the secondary centre of diversity of the subfamily (Langenheim *et al.*, 1973). At least 26 Caesalpinioideae genera are endemic to tropical Asia and 12 genera to tropical Asia and Africa (Lee, 1976). Members of the Leguminosae are well represented in various forest types in India, with several important species, both economically and ecologically. There are at least 1,192 species of Leguminosae belonging to 173 genera found in India (Arisdason and Lakshminarasimhan, 2016), indicating the vast richness of the group in the continent. Peninsular India represents the highest diversity of legumes with more than 550 species whereas the Himalayas have the second highest numbers with 500 species (Hussain and Kopoor, 1990). On the other hand, more than 400 legume species are reported at different altitudinal gradients and forest types in northeast India, which falls in the eastern Himalayan region (Sanjappa, 1992). A summary of legume species occurrence in the Indian subcontinent is given in Fig. 2.1.

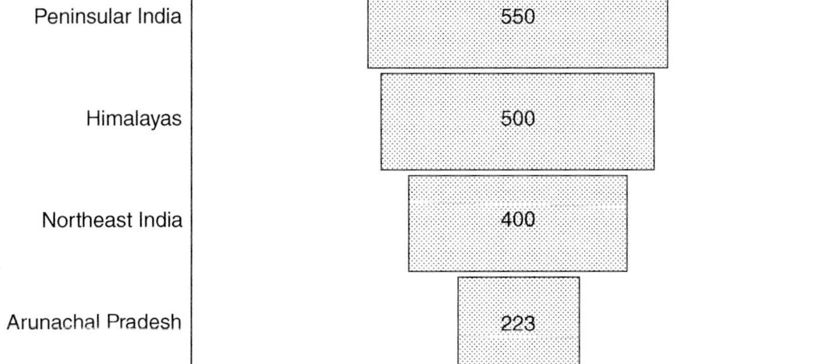

Fig. 2.1. Funnel chart showing legume species richness in the Indian subcontinent (Sanjappa, 1992; Sharma *et al.*, 2002; Arisdason and Lakshminarasimhan, 2016).

Though legume taxonomy in India is well studied, the northeastern states, especially Arunachal Pradesh, remain less explored due to remoteness, rugged terrain and extreme weather conditions. A few studies have showed the legume diversity of the eastern Himalayan region in general and Arunachal Pradesh in particular and documented a few key species of multiple utilization opportunities (Dash and Singh, 1998; Dash, 2009; Kalita *et al.*, 2015). The edapho-climatic conditions of northeastern India favour luxuriant growth of vegetation attaining the stage of climax formation which is represented by the scattered oak forests of the region. The region supports luxuriant growth of enormous diversity of plant resources including several primitive taxa and is considered to be the 'cradle of flowering plants' (Takhtajan, 1969). The state also has a few rare and endemic legume taxa such as *Albizia arunachalensis*, *Albizia gamblei*, *Crotalaria bhutanica* and *Dalbergia clarkei*. Sharma *et al.* (2002) accounted a total of 223 species of legumes belonging to 73 genera in Arunachal Pradesh. Leguminosae is one of the dominant plant groups in the eastern Himalayan region and are being utilized for different sociocultural and economic purposes (Chowdhery *et al.*, 1996; Dash, 2009). However, no systematic phylogenetic studies have been conducted for legumes of the eastern Himalayan region.

2.3 *Gymnocladus* and its Affinity

Gymnocladus and its sister genus *Gleditsia* consist of very few species but they are distributed in a broad geographic range in Asia, Africa and North and South America. Caesalpinioideae is generally distributed pantropically and common in both wet and dry regions. However, the distribution range of the genera *Gleditsia* and *Gymnocladus* extends up to the temperate zones as well (Azani *et al.*, 2017). *Gymnocladus* is a small archaic genus consisting of just a handful of species belonging to the subfamily Caesalpinioideae. The genus *Gymnocladus* was first reported in North America and Africa with only two species, *Gymnocladus canadensis* and *Gymnocladus arabicus*. A few other species of *Gymnocladus* were described later and reviewed multiple times. *Gymnocladus dioica* (L.) K. Koch (Fig. 2.2) is widely distributed in North America. Altogether, five *Gymnocladus* species are taxonomically accepted as per International Legume Database and Information Service (ILDIS) parameters (Roskov *et al.*, 2005). While two species, namely *G. dioica* (L.) K. Koch and *G. angustifolius* (Gagnep.) J.E. Vidal are restricted to North America and Vietnam respectively, *G. chinensis* Baill., *G. assamicus* Kanj. ex P.C. Kanjilal and *G. burmanicus* C.E. Parkinson are confined to the trijunction of India, China and Myanmar. *G. assamicus* was first collected in 1920 from the Khasi Hills in Meghalaya (Laitkshe, Marngor, 1,800 m) (Kanjilal *et al.*, 1934). *G. assamicus* remained unnoticed, mainly due to its extremely rare population despite several attempts by various researchers. *Gymnocladus burmanicus* is a poorly documented species distributed at about 1,000 m above sea level. It was documented from East Siang district in Arunachal Pradesh (India) during extensive field explorations. Fully mature trees attain around 40–50 m in hight, having a straight trunk and bright green canopy (Fig. 2.3). Rachis of *G. burmanicus* leaves are around 20 cm long, sparsely pubescent to glabrate, usually with three pairs of pinnae. Leaflets

Fig. 2.2. Mature *Gymnocladus dioica* trees planted roadside across North America (photo: Baharul I. Choudhury).

Fig. 2.3. A young *G. burmanicus* tree in East Siang district of Arunachal Pradesh (India) (photo: Baharul I. Choudhury).

are 4–6 cm long, ovate-lanceolate and oblique with narrow bases and apex of all leaflets acute to acuminate (Fig. 2.4). A summary of geographic distribution of the *Gymnocladus* species is given in Table 2.1.

Another sister genus of *Gymnocladus* is *Gleditsia* which was described by Linnaeus with a single species *Gleditsia triacanthos* from Virginia (North America). Currently, there are about 14 species in the world out of which 10 are found in temperate Asia, and 2 or 3 in North America. *G. amorphoides* (Griseb.) Taub. is found in South America and *G. caspica* Desf. in the Caspian Sea region. In India, the genus *Gleditsia* is represented by *Gleditsia assamica* Bor. and *Gleditsia amorphoides* (Griseb.) Taub (Roskov *et al.*, 2005). *Gleditsia assamica* is also restricted to the northeastern region of India (Sanjappa, 2002) and remnant populations were documented in the Nokrek Biosphere Reserve, India (Fig. 2.5). A recent account predicted the occurrence of about 16 species of modern *Gleditsia* widely distributed in central and southeast Asia as well as in North and South America and three or four species of *Gymnocladus* distributed in North America and South Asia (Chen *et al.*, 2010).

Fig. 2.4. Leaflets of *G. burmanicus* (photo: Baharul I. Choudhury).

Table 2.1. Worldwide distribution of *Gymnocladus* species.

Sl. No.	Species	Distribution
1	*Gymnocladus angustifolius*	Asia: Vietnam
2	*Gymnocladus assamicus*	Asia: India: northeast India
3	*Gymnocladus burmanicus*	Asia: India, Myanmar
4	*Gymnocladus chinensis*	Asia: China, India
5	*Gymnocladus dioica*	Asia, Europe, North America, Pacific Ocean, South America

2.4 Geographic Distribution of *Gymnocladus* and *Gleditsia* Species

Gymnocladus was first described by Lamark, with two species *Gymnocladus canadensis* and *Gymnocladus arabicus* from eastern Canada and Egypt respectively. *Guilandina dioca* L. is an earlier name for the Canadian species *Gymnocladus canadensis* which is again synonymized to *Gymnocladus dioica* (L.) K. Koch (Lee, 1976). Lamark's other species *G. arabicus* is now considered synonymous with *Moringa optera* of the family Moringaceae (Sanjappa, 2002).

The genus *Gymnocladus* shows interesting patterns of global distribution. According to the ILDIS (Roskov *et al.*, 2005), five species of *Gymnocladus* are taxonomically accepted. However, according to Missouri Botanical Garden (TROPICOS Nomenclatural Database) nine species of *Gymnocladus* are distributed worldwide (Anonymous, 2018). Three species of *Gymnocladus*, namely *G. dioica* (L.) K. Koch, *G. arabicus* Lam. and *G. angustifolius* (Gagnep.) J.E. Vidal are reported to be restricted to North America, Egypt and Vietnam respectively while the remaining three species (*G. chinensis* Baill., *G. assamicus* Kanj. ex P.C. Kanjilal and *G. burmanicus* C.E. Parkinson) are confined to the trijunction of India, China and Myanmar. In India, *Gymnocladus* is represented by the occurrence of *G. assamicus* *G. burmanicus* and *G. chinensis*, known only in the northeastern states (Sanjappa,

Fig. 2.5. A mature *Gleditsia assamica* tree in Nokrek Biosphere Reserve, Meghalaya (India) (photo: Baharul I. Choudhury).

2002). Though the genus *Gymnocladus* has worldwide distribution, *G. assamicus* is endemic to northeast India and is mainly distributed in specific pockets of high altitude areas in Arunachal Pradesh (Choudhury *et al.*, 2007). *G. assamicus* is also distributed in the Khasi hills in Meghalaya (type locality) (Kanjilal *et al.*, 1934; Venugopal and Pamidimarri, 2015).

2.5 Phylogenetic Relationship

The subfamily Caesalpinioideae occupies a basal position in phylogenetic schemes and is highly diverse in floral form and ontogeny (Doyle *et al.*, 2000; Bruneau *et al.*, 2001). Several members of Caesalpinioideae display primitive forms of reproductive and vegetative organs, highlighting their old ancestry in the phylogenetic tree. The Caesalpinioideae are usually divided into four or five tribes based on traditional morphology-based systems of classification and are considered as one of the most poorly understood groups of angiosperm from a phylogenetic perspective. Caesalpinioideae constitutes a paraphyletic assemblage of taxa and shows complex phylogenetic relationships. At the same time, subfamilies Mimosoideae and Papilionoideae, which are derived from the Caesalpinioideae, are considered monophyletic (Kass and Wink, 1996; Pennington *et al.*, 2000).

Several studies based on morphological data (Lewis and Schrire, 1995), molecular data (Simpson and Miao, 1997; Bruneau *et al.*, 2001; Gervais and Bruneau, 2002) and a combination of both morphological and molecular data (Herendeen *et al.*, 2003a) have been used to resolve the complex phylogenetic relationships of the group. Distribution of *Gymnocladus* and *Gleditsia* across temperate Asia and North America has attracted many researchers to study the phylogeographic patterns between the two continents. They have found remarkable similarities between temperate East Asian and North American flora that have been recognized since the time of Linnaeus (Graham, 1972; Boufford and Spongberg, 1983; Tiffney, 1985).

2.5.1 The unique Umtiza clade

The Umtiza clade forms a unique group with *Gymnocladus*, *Gleditsia*, *Umtiza*, *Tetrapterocarpon*, *Arcoa*, *Acrocarpus* and *Ceratonia* (Lewis, 2005). This small clade is important from taxonomic, biogeographic and evolutionary perspectives. These seven genera show notable differences in terms of their habitat, life form and several other morphological variations. In general, Caesalpinioideae is rarely distributed in the temperate zone but, interestingly, many species of the Umtiza clade are distributed in temperate regions as well (Fig. 2.6). More importantly, the group has been described as dioecious and is the main reason for showing affinities to each other. Flowers in the Umtiza clade are small, pedicellate, actinomorphic and tubular (Fig. 2.7). Individual flowers are narrowed at the base and gradually widen at the mouth. The corolla tube is 5–10 mm long and 3–4 mm wide with a shaggy,

Fig. 2.6. Distribution of different *Gymnocladus* species throughout the world.

Fig. 2.7. A representative inflorescence of Umtiza clade (*Gymnocladus assamicus*) showing small, pedicellate, actinomorphic and tubular flowers (photo: Baharul I. Choudhury).

hairy outside. Stamens are didynamous, alternate with long and short filaments (Fig. 2.8). Anthers are oblong, ovoid and bithacous. Flower colour varies a little from greenish-white to purple.

Gleditsia triacanthos flowers are greenish-yellow in colour, inflorescence-type fascicled racemes that are sometimes borne directly from the tree trunk (Fig. 2.9). Flowers of *Gymnocladus dioica* are white with green perianth and are poorly differentiated (Fig. 2.10). However, *Gymnocladus assamicus* flowers are much

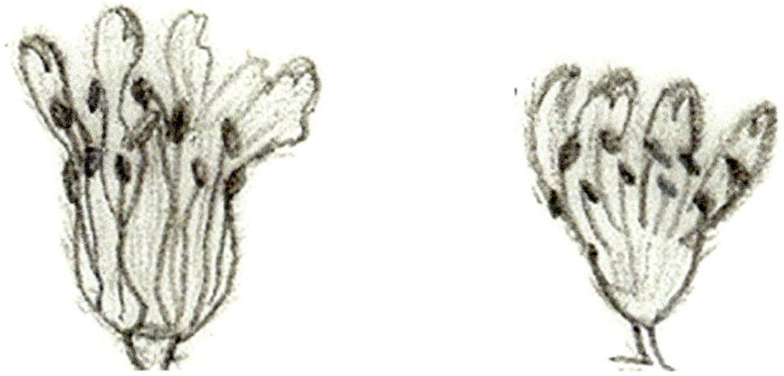

Fig. 2.8. Hermaphrodite (left) and male *G. assamicus* flowers showing didynamous stamens.

Fig. 2.9. Greenish-yellow *Gleditsia triacanthos* flowers borne directly from the tree trunk (photo: Baharul I. Choudhury).

more showy with a brilliant purple colour but still display the primitive traits similar to other members of the Umtiza clade (Fig. 2.11).

Perianth structure is a striking and primitive character of the Umtiza clade. The perianths in these genera fused or poorly differentiated into clayx and corolla and are of the same colour. Male and hermaphrodite flowers of *Gymnocladus assamicus*, *Gymnocladus dioica* and *Gleditsia triacanthos* are shown in Fig. 2.12. Greenish and undifferentiated perianths in the Umtiza clade are examples of apomorphies or derived traits and not plesiomorphic (ancestral character)

Fig. 2.10. *Gymnocladus dioica* male inflorescence with white flowers (photo: Baharul I. Choudhury).

Fig. 2.11. *Gymnocladus assamicus* inflorescence showing brilliant purple flowers (photo: Baharul I. Choudhury).

features for the family (Herendeen *et al.*, 2003a). The major morphological traits that may be attributed to the clade are habit of the species, flower colour, life forms and, most importantly, floral sexuality. However, there are instances where apparently bisexual flowers were observed in multiple genera and it is difficult to ascertain accurate evolutionary history of sexual systems in the Umtiza clade (Herendeen *et al.*, 2003b).

Subfamily Caesalpinioideae is generally divided into five tribes, Caesalpinieae, Cassieae, Cercideae, Detarieae and Amherstieae (Polhill, 1994). Detarieae is the largest tribe within Caesalpinioideae and is monophyletic, excluding the monotypic South African genus *Umtiza* (Polhill, 1994; Bruneau et al., 2001; Herendeen et al., 2003a). Placement of *Umtiza* in the tribe Detarieae has been debated for a long time due to its unique and rare traits such as absence of stipules (Léonard, 1957; Banks and Klitgaard, 2000; Herendeen *et al.*, 2003a). In general, 'intrapetiolar' insertion of the stipules is a key identifying feature of the tribe Detarieae (Polhill and Vidal, 1981). The placement of *Umtiza listeriana* in Detarieae is primarily based on identical corolla, however, the calyx with free sepals is shorter and does not cover the corolla (Léonard, 1957). *Umtiza* also has an unusual cyme inflorescence unlike a typical raceme common in Caesalpinioideae (Tucker, 2001). Palynological studies showed different pollen structures in *Umtiza* to the other species of Detarieae (Ferguson and Banks, 1995; Banks and Klitgaard, 2000) and attempts have been made to remove *Umtiza* from the Detarieae and place them in more suitable clades in the Caesalpinioideae (Léonard, 1957). However, a phylogenetic study using only molecular data could not resolve the placement of *Umtiza* either (Bruneau *et al.*, 2001).

2.6 Placement of *Gymnocladus* and *Gleditsia* in the Phylogenetic Tree

A phylogenetic study of *Gymnocladus* and *Gleditsia* by Schnabel *et al.* (2003) revealed their centre of origin to be eastern Asia, originating during the Eocene. Eastern North American species of both the genera are most likely to be evolved from ancestors that migrated across the Bering land bridge, but the eastern Asian/eastern North American disjunction appeared to be more primitive in *Gymnocladus* than in *Gleditsia* (Schnabel *et al.*, 2003). This is also an example showing disjunctions between the northern and southern hemispheres of the temperate legume lineages; however, instances are available showing a rise of tropical clades from the temperate legumes (Pennington and Dick, 2010). In a comprehensive study using combined morphological and molecular data, Herendeen *et al.* (2003a) placed genera such as *Ceratonia*, *Acrocarpus* and *Tetrapterocarpon* along with *Gleditsia* and *Gymnocladus* in the *Umtiza* clade. In particular, the genera *Gymnocladus* and *Gleditsia* show a distinctive clade along with *Umtiza* which is a monospecific genus and possesses many potential morphological synapomorphies (Herendeen *et al.*, 2003b). One of the most striking characteristics of the clade is dioecy, in which distinct male, female and/or hermaphrodite individuals coexist in the same population.

Gymnocladus assamicus (male and hermaphrodite flower)

Gymnocladus dioica (male and hermaphrodite flower)

Gleditsia triacanthos (male and hermaphrodite flower)

Fig. 2.12. Flowers of Umtiza clade species (*Gymnocladus* and *Gleditsia*) showing sexual dimorphism (photo: Baharul I. Choudhury).

Dioecy is usually common in animals (especially vertebrates) but uncommon or rare in flowering plants (Smith and Maynard-Smith, 1978). In plants, dioecious species are enumerated in 6% (14,620) of the species, in 7.1% (959) of genera and 38% (167) of families (Renner and Ricklefs, 1995).

Inclusion of genera such as *Gleditsia* and *Gymnocladus* added a few extra features including more diverse geographic distribution and morphological traits. For example, *Umtiza* possess cyme inflorescence which is also a characteristic of the *Poeppigia*, *Gleditsia*, *Ceratonia* and genera of Dialiinae of Cassieae (Tucker, 1998, 2001). Floral morphology of *Gymnocladus* and *Gleditsia* show primitive characteristics such as greenish perianth which are not well differentiated (Herendeen *et al.*, 2003b). *Gleditsia* also possesses other primitive characteristics such as unstable floral structure, polygamy, radial floral symmetry and unspecialized pollen (Gordon, 1966; Polhill, 1981; Polhill and Vidal, 1981). Tucker (1991) showed multiple evidences of primitive floral developmental features by comparing six *Gleditsia* species, such as unequal-sized flower primordia which produce flowers of variable merosity; occurrence of floral bracts; variable number of organs; and position of the first organ (sepal) initiated. The raceme inflorescence type often bears lateral branches near the base, or fascicles of flowers at some points (Tucker, 1991). *Gleditsia triacanthos* show phenotypic instability in which thornless trees turn thorn-bearing after 40 to 55 years, demonstrating that the thornless phenotype is not genetically fixed or a simple recessive trait (Michener, 1986).

Gymnocladus species show unique geographic distribution from eastern Asia to North America and it is challenging to study the phylogeny extensively. On the basis of available phylogenetic studies, a hypothetical phylogenetic tree has been created to show the most probable phylogenetic relationships between *Gymnocladus*, *Gelditsia* and *Umtiza* species (Fig. 2.13). However, more detailed study is needed to understand the complex evolutionary history of the Umtiza clade in general and the geneus *Gymnocladus* in particular.

2.7 Complex Evolutionary History of *Gymnocladus* and *Gleditsia*

The floral characteristics of the genus *Gymnocladus* are bedevilling. Some taxonomic accounts have described the genus *Gymnocladus* as polygamous or dioecious (Chen *et al.*, 2010) while others have described the genus with unisexual flowers (Watson and Dallwitz, 1993). On the other hand, *Gymnocladus chinensis* has been described as polygamous (Jia, 1875). The genus *Gymnocladus* also shows similar morphological traits such as dioecious nature, bipinnate leaves, diplostemonous androecium, poorly differentiated greenish calyx and corolla. However, *Gymnocladus* trees do not bear spines and the pods are fleshy, saponaceous and dehiscent. Based on such primitive characteristics, *Gymnocladus*, *Gleditsia* and *Umtiza* are considered archaic members in the family (Herendeen *et al.*, 2003a). There are also remarkable similarities in wood anatomy of *Gymnocladus* and *Gleditsia*, with fossil wood from miocene angiosperms (Cheng *et al.*, 2018) highlighting their primitive nature. Detailed

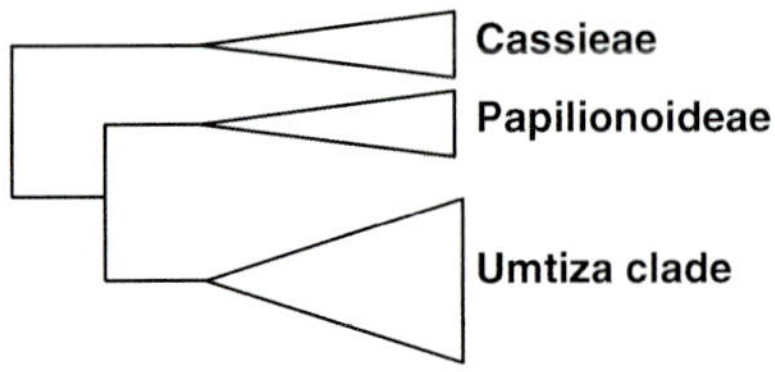

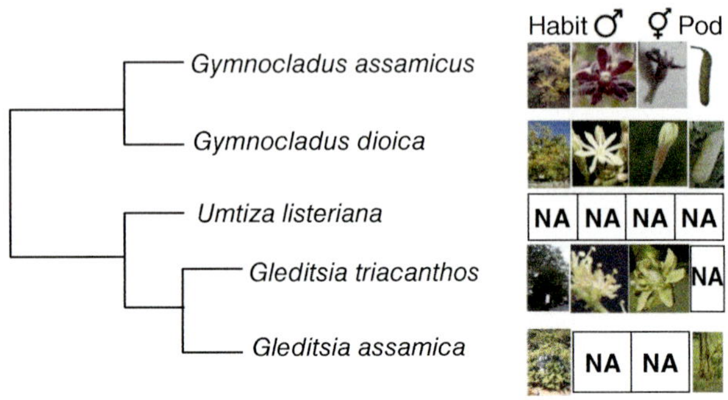

Fig. 2.13. Hypothetical phylogenetic relationships among Umtiza clade species based on published data (Herendeen *et al.*, 2003a; Schnabel *et al.*, 2003; Lewis, 2005).

study of life history traits, reproductive behaviour and paternity tests of multiple Umtiza clade species will be helpful in resolving the species dilemma in terms of phylogenetic relationships.

2.8 References

Anonymous, 2018. *TROPICOS* – Tropicos.org. Missouri Botanical Garden. 06 Nov 2018, accessed 15 December 2018, available at URL: http://www.tropicos.org/Name/40018822?tab=subordinatetaxa.

Arisdason, W. and Lakshminarasimhan, P., 2016. *Status of Plant Diversity in India: An Overview.* Central National Herbarium, Botanical Survey of India, Howrah.

Azani, N., Babineau, M., Bailey, C.D., Banks, H., Barbosa, A.R., Pinto, R.B. and Candido, E., 2017. A new subfamily classification of the Leguminosae based on a taxonomically comprehensive phylogeny. The Legume Phylogeny Working Group (LPWG). *Taxon* 66.1: 44–77.

Bandel, G., 1974. Chromosome numbers and evolution in the Leguminosae. *Caryologia,* 27(1): 17–32.

Banks, H. and Klitgaard, B.B., 2000. Palynological contribution to the systematics of detarioid legumes (Leguminosae: Caesalpinioideae). In P.S. Herendeen and A. Bruneau (eds) *Advances in Legume Systematics,* pp. 79–106 (Pt 9). Royal Botanic Gardens, Kew, London.

Bouchenak-Khelladi, Y., Maurin, O., Hurter, J. and Van der Bank, M., 2010. The evolutionary history and biogeography of Mimosoideae (Leguminosae): an emphasis on African acacias. *Molecular Phylogenetics and Evolution* 57(2): 495–508.

Boufford, D.E. and Spongberg, S.A., 1983. Eastern Asian-eastern North American phytogeographical relationships – a history from the time of Linnaeus to the twentieth century. *Annals of the Missouri Botanical Garden*, Vol. 70, No. 3: 423–439.

Bruneau, A., Forest, F., Herendeen, P.S., Klitgaard, B.B. and Lewis, G.P., 2001. Phylogenetic relationships in the Caesalpinioideae (Leguminosae) as inferred from chloroplast trnL intron sequences. *Systematic Botany* 26(3): 487–514.

Bruneau, A., Mercure, M., Lewis, G.P. and Herendeen, P.S., 2008. Phylogenetic patterns and diversification in the caesalpinioid legumes. *Botany* 86(7): 697–718.

Chen, D., Zhang, D. and Larsen, K., 2010. GYMNOCLADUS Lamarck, Encycl., 1: 733. 1785, nom. Cons. Flora of China 10:36, accessed 14 December 2018, http://flora.huh.harvard.edu/china/PDF/PDF10/Gymnocladus.pdf.

Chen, D.Z., Zhang, D.X., Ding, H., Larsen, K., Larsen, S.S., 2010. Tribe Caesalpinieae (Fabaceae). In Z.Y. Wu, P.H. Raven and D.Y. Hong (eds) *Flora of China*, 10: 35–49. Missouri Botanical Garden Press, St. Louis, MO.

Cheng, Y.M., Yang, X.N., He, Z.F., Mao, B. and Yin, Y.F., 2018. Early Miocene angiosperm woods from Sihong in the Jiangsu Province, Eastern China. *IAWA Journal*, 39(01): 125–142.

Choudhury, B.I., Khan, M.L., Arunachalam, A. and Das, A.K., 2007. Ecology and conservation of the critically endangered tree species *Gymnocladus assamicus* in Arunachal Pradesh, India. *International Journal of Ecology*, 2007, ID 59282.

Chowdhery, H.J., Giri, G.S., Pal, G.D., Pramanik, A. and Das, S.K., 1996. Fabaceae, Caesalpiniaceae and Mimosaceae. In P.K. Hajra, B.M. Burma and G.S. Giri (eds) *Materials for the Flora of Arunachal Pradesh*, Vol. 1. BSI, Brabourne Road, Kolkata, India, pp. 345–414.

Dash, S.S., 2009. The genus *Desmodium* Desv. (Liguminosae) and its allied genera in Arunachal Pradesh. *Bulletin of Arunachal Forest Research* 25(1&2): 18–33.

Dash, S.S. and Singh, P., 1998. The genus *Desmodium* Desv. (Liguminosae) in Sikkim. *Journal of Hill Research* 11(2): 212–223.

Doyle, J.J. and Luckow, M.A., 2003. The rest of the iceberg. Legume diversity and evolution in a phylogenetic context. *Plant Physiology* 131(3): 900–910.

Doyle, J.J., Chappill, J.A., Bailey, C.D. and Kajita, T., 2000. Towards a comprehensive phylogeny of legumes: evidence from rbcL sequences and non-molecular data. *Advances in Legume Systematics*, 9: 1–20.

Ferguson, I.K. and Banks, H., 1995. Pollen morphology of the *Cynometra* group (Detarieae: Caesalpinioideae: Leguminosae). In N.M. Dutta, S. Gupta-Bhattacharya and M. Bhattacharya (eds) S*econd Symposium of African Palynology, Tervuren* (Belgium). Publ Occas CIFEG 1995/31, Orleans, pp. 217–227.

Gervais, G.Y.F. and Bruneau, A., 2002. Phylogenetic analysis of a polyphyletic African genus of Caesalpinioideae (Leguminosae): Monopetalanthus Harms. *Plant Systematics and Evolution*, 235(1–4): 19–34.

Gordon, D. 1966. A revision of the genus *Gleditsia* (Leguminosae). PhD dissertation, Indiana University, Bloomington, IN.

Graham, A., 1972. Outline of the origin and historical recognition of floristic affinities between Asia and eastern North America. *Floristics and Paleoflorists of Asia and Eastern North America*, 15.

Graham, P.H. and Vance, C.P., 2003. Legumes: importance and constraints to greater use. *Plant Physiology*, 131(3): 872–877.

Herendeen, P.S., Crepet, W.L., Dilcher, D.L., Herendeen, P.S. and Dilcher, D.L., 1992. The fossil history of the Leguminosae: phylogenetic and biogeographic implications.

Advances in Legume Systematics, part 4, The Fossil Record. Royal Botanic Gardens Kew, UK, pp. 303–316.

Herendeen, P.S., Bruneau, A. and Lewis, G.P., 2003a. Phylogenetic relationships in caesalpinioid legumes: a preliminary analysis based on morphological and molecular data. *Advances in Legume Systematics*, Part 10, pp. 37–62.

Herendeen, P.S., Lewis, G.P. and Bruneau, A., 2003b. Floral morphology in caesalpinioid legumes: testing the monophyly of the 'Umtiza clade'. *International Journal of Plant Sciences*, 164(S5): S393–S407.

Hussain, T. and Kopoor, S.L., 1990. *Enumeration of Legumes of India: Indigenous and Introduced.* NBRI, Lucknow, India, pp. 4–61.

Jia, F.Z., 1875. *Gymnocladus chinensis* Baillon. *Bulletin Mensuel de la Societe Linneenne de Paris* 1(34).

Kalita, B.C., Bharali, P., Jamoh, L. and Tag, H., 2015. Diversity of legumes in the Papum Pare District of Arunachal Pradesh of India. *Pleione* 9(2): 440–448.

Kanjilal, U.N., Kanjilal, P.C., Das, A., Purkayastha, C. and Bor, N.L., 1934. A brief ecological sketch of the botany of Assam. *Flora of Assam* 1(a), Government Press, Shillong, India.

Kass, E., and Wink, M., 1996. Molecular evolution of the Leguminosae: phylogeny of the three subfamilies based on rbcL-sequences. *Biochemical Systematics and Ecology* 24(5): 365–378.

Langenheim, J.H., Lee, Y.T. and Martin, S.S., 1973. An evolutionary and ecological perspective of Amazonian Hylaea species of Hymeneae (Leguminosae: Caesalpinioideae). *Acta Amazonica* 3(1): 5–38.

Lavin, M., Herendeen, P.S. and Wojciechowski, M.F., 2005. Evolutionary rates analysis of Leguminosae implicates a rapid diversification of lineages during the tertiary. *Systematic Biology* 54(4): 575–594.

Lee, Y.T., 1976. The genus Gymnocladus and its tropical affinity. *Journal of the Arnold Arboretum*, 57(1): 91–112.

Léonard, J., 1957. Genera des Cynometreae et des Amherstieae africaines (Leguminosae-Cacsalpinioidcac). Essai dc blastogénie appliquée à la systématique. *Mémie Acadale royde Belgique. Classedes Sciences. Mémoires*, 8vo 30(2): 1–314.

Lewis, G.P., 2005. Tribe Caesalpinieae. In G.P. Lewis, B. Schrire, B. Mackinder and M. Lock (eds) *Legumes of the World*, Royal Botanic Gardens, Kew, pp. 127–161.

Lewis, G.P. and Schrire, B.D., 1995. A reappraisal of the Caesalpinia group (Caesalpinioideae: Caesalpinieae) using phylogenetic analysis. *Advances in Legume Systematics: Part Seven. Phylogeny.* Royal Botanic Gardens, Kew, pp. 41–52.

Lewis, G.P., Schrire, B., Mackinder, B. and Lock, M. (eds), 2005. *Legumes of the World.* Royal Botanic Gardens, Kew.

Manzanilla, V. and Bruneau, A., 2012. Phylogeny reconstruction in the Caesalpinieae grade (Leguminosae) based on duplicated copies of the sucrose synthase gene and plastid markers. *Molecular Phylogenetics and Evolution* 65(1): 149–162.

Michener, D.C., 1986. Phenotypic instability in Gleditsia triacanthos (Fabaceae). *Brittonia* 38(4): 360–361.

Morley, R.J., 2000. *Origin and Evolution of Tropical Rain Forests.* John Wiley & Sons, New York.

Pennington, R.T. and Dick, C.W., 2010. *Diversification of the Amazonian Flora and its Relation to Key Geological and Environmental Events: A Molecular Perspective.* Blackwell, Oxford.

Pennington, R.T., Klitgaard, B.B., Ireland, H.E.L.E.N and Lavin, M., 2000. New insights into floral evolution of basal Papilionoideae from molecular phylogenies. *Advances in Legume Systematics* 9: 233–248.

Polhill, R.M., 1981. Evolution and systematics of the Leguminosae. *Advances in Legume Systematics* 1, Royal Botanic Gardens, Kew, pp.1–26.

Polhill, R.M., 1994. Complete synopsis of legume genera. In F.A. Bisby, J. Buckingham and J.B. Harborne (eds) *Phytochemical Dictionary of the Leguminosae, Vol. 1, Plants and their Constituents*. Chapman & Hall, London, pp. xlix–liv.

Polhill, R.M., and Vidal, J.E., 1981. Caesalpinieae. In R.M. Polhill and P.H. Raven (eds) *Advances in Legume Systematics*. Royal Botanic Gardens, Kew, pp. 81–96.

Ramdhani, S., Cowling, R.M. and Barker, N.P., 2010. Phylogeography of Schotia (Fabaceae): recent evolutionary processes in an ancient thicket biome lineage. *International Journal of Plant Sciences*, 171(6): 626–640.

Raven, P.H. and Axelrod, D.I., 1974. Angiosperm biogeography and past continental movements. *Annals of the Missouri Botanical Garden* 61(3): 539–673.

Renner, S.S. and Ricklefs, R.E., 1995. Dioecy and its correlates in the flowering plants. *American Journal of Botany* 82(5): 596–606.

Roskov, Y.R, Bisby, F.A., Zarucchi, J.L., Schrire, B.D. and White, R.J. (eds), 2005. *ILDIS World Database of Legumes* (draft checklist, version 10). CD-ROM: ILDIS, University of Reading, Reading, UK.

Sanjappa, M., 1992. *Legumes of India*. Bishen Singh & Mahendra Pal Singh, Dehradun, India, pp. 6–82.

Sanjappa, M., 2002. *Gleditsia* and *Gymnocladus* Lam.(Leguminosae-Caesalpiniodeae) in India. In R.R. Rao and L.B. Chaudhary (eds) *Advances in Legume Research in India*. Bishen Singh Mahendra Pal Singh, Dehradun, India, pp. 27–34.

Schnabel, A., McDonel, P.E. and Wendel, J.F., 2003. Phylogenetic relationships in *Gleditsia* (Leguminosae) based on ITS sequences. *American Journal of Botany* 90(2): 310–320.

Schrire, B.D., Lavin, M. and Gwilym, P.L., 2005a. Plant diversity and complexity patterns: local, regional and global dimensions. *Proceedings of an International Symposium Held at the Royal Danish Academy of Sciences and Letters in Copenhagen*, Denmark, 25–28 May 2003, 55. Kgl. Danske Videnskabernes Selskab.

Schrire, B.D., Lewis, G.P. and Lavin, M., 2005b. Biogeography of the Leguminosae. In G. Lewis, B.D. Schrire, B. Mackinder and M. Lock (eds) *Legumes of the World*. Royal Botanic Gardens, Kew, pp. 21–54.

Sharma, A., Haridasan, K. and Borthakur, S.K., 2002. Notes on the Legume flora of Arunachal Pradesh with special reference to West Kameng district. In R.R. Rao (ed.) *Advances in Legume Research in India*. Bisen Singh Mahendra Pal Singh, Dehradun, India, pp. 171–179.

Simpson, B.B. and Miao, B.M., 1997. The circumscription of *Hoffmannseggia* (Fabaceae, Caesalpinioideae, Caesalpinieae) and its allies using morphological and cpDNA restriction site data. *Plant Systematics and Evolution* 205(3–4): 157–178.

Smith, J.M. and Maynard-Smith, J., 1978. *The Evolution of Sex* (No. 574.1 S5). Cambridge University Press, Cambridge.

Sprent, J.I., 2007. Evolving ideas of legume evolution and diversity: a taxonomic perspective on the occurrence of nodulation. *New Phytologist* 174(1): 11–25.

Takhtajan, A.L., 1969. *Flowering Plants: Origin and Dispersal*. Smithsonian Institution Press, Washington, DC, 310 pp.

Tiffney, B.H., 1985. Perspectives on the origin of the floristic similarity between eastern Asia and eastern North America. *Journal of the Arnold Arboretum* 66(1): 73–94.

Tucker, S.C., 1991. Helical floral organogenesis in *Gleditsia*, a primitive caesalpinioid legume. *American Journal of Botany* 78(8): 1130–1149.

Tucker, S.C., 1998. Floral ontogeny in legume genera *Petalostylis*, *Labichea*, and *Dialium* (Caesalpinioideae: Cassieae), a series in floral reduction. *American Journal of Botany* 85: 184–208.

Tucker, S.C., 2001. Organ loss in detarioid and other leguminous flowers, and the possibility of saltatory evolution. *Advances in Legume Systematics:* Part 9: 107–120.
Tucker, S.C., 2003. Floral development in legumes. *Plant Physiology*, 131(3): 911–926.
Venugopal, N. and Pamidimarri, S., 2015. Rediscovery of critically endangered *Gymnocladus assamicus* Kanjilal ex PC Kanjilal (Leguminosae: Caesalpinioideae) from West Khasi Hills District of Meghalaya. *Pleione* 9: 186–192.
Watson, L. and Dallwitz, M.J. 1993 onwards. The genera of Leguminosae-Caesalpinioideae and Swartzieae: descriptions, illustrations, identification, and information retrieval. In English and French. Version: 10th May 2019. https://www.delta-intkey.com/caes/en/www/humboldt.htm, accessed 26 June 2019.
Wing, S. L., Herrera, F. and Jaramillo, C., 2004. A Paleocene flora from the Cerrajón Formation, Guajíra Peninsula, northeastern Colombia. *VII International Organization of Paleobotany Conference Abstracts*, Trelew, Argentina. Museo Egidio Feruglio, Argentina, pp.146–147.
Yahara, T., Javadi, F., Onoda, Y., de Queiroz, L.P., Faith, D.P., Prado, D.E., Akasaka, M., Kadoya, T., Ishihama, F., Davies, S. and Slik, J.W., 2013. Global legume diversity assessment: concepts, key indicators, and strategies. *Taxon*, 62(2): pp. 249–266.

3 Reproductive Ecology of *Gymnocladus assamicus*

3.1 Introduction

Plant reproductive ecology is one of the most rapidly developing fields in plant science and gives enormous knowledge on breeding systems, reproductive lifespan and flower, fruit and seed production of a species. The processes of gamete development, endosperm and embryo development and other reproductive features, which are crucial for continuity of a plant life cycle, are demonstrated from such studies and contribute vital information about the biology of a species. Detailed study of breeding systems of a species is vital to understand different life history strategies, reproductive success as well as interaction with different biotic and abiotic components. Reproductive biology studies at the community level also help to understand the mechanisms of gene flow and speciation in plant communities (Baker, 1959; Bawa, 1980; Baack *et al.*, 2015; Pinheiro *et al.*, 2015; Kua and Cannon, 2017).

Reproductive ecology studies also help in developing strategies for conservation of species, especially rare and endangered taxa, and can provide important clues regarding the reproductive constraints of such species (Carlsen *et al.*, 2002; Bailey and Kevan, 2017). Such investigations help in developing certain protocols to combat the problems that impede regeneration, especially for species which bear fewer propagules for future generations. Studies on reproductive ecology also help in developing strategies to preserve the genetic potential of rare/endangered/threatened species and are crucial for restoration and reintroduction (Moza and Bhatnagar, 2007).

Plants often develop various reproductive strategies by acquiring different characteristics in terms of flower development, fruit and seed set and dispersal. Angiosperms have developed striking variations in reproductive systems which in turn play major roles in lineage diversification and evolutionary processes (Barrett,

2013). Flowering plants can reproduce either asexually and apomixis or sexually, or both, to produce genetically identical offspring and continue the next generation. Diverse sexual and asexual reproductive mechanisms and strategies have been evolved in flowering plants to adapt to suitable and respective environments. Sexual reproduction is also associated with a variety of pollination and mating systems which are effective ways of exchanging genetic material or genes essential for population survival of most flowering plants. If gene flow between plant populations is low, it will cause inbreeding and poor adaptation of the subpopulations to the changing environment, eventually leading to extinction (Rasmussen and Kollmann, 2004). Mechanisms of sexual reproduction are also found to be more advantageous in the wake of future climate change scenarios (Crow, 1992). Studies also suggest changes in phenology and pollinator interaction of different plant species in response to climate change (Hegland *et al.*, 2009; Willmer, 2012; Forrest 2015).

Population size is an important parameter to maintain a genetically healthy population. Large populations have the ability to maintain gene flow and genetic variation while smaller populations are prone to extinction due to low genetic diversity. Several factors such as inbreeding, genetic drift, restricted gene flow and small population size have been directly implicated in the reduction of genetic diversity. In particular, rare and endangered species with fragmented populations are prone to these consequences and likely early extinction (Saccheri *et al.*, 1998; Frankham *et al.*, 2010). Evidence suggests that populations having low genetic diversity also exhibit increased rates of extinction (Markert *et al.*, 2010).

Populations of rare or endangered plant species are characterized by very few actively reproducing individuals, fragmented populations and lower percentage of seedling or sapling populations compared to mature trees (Choudhury *et al.*, 2007). Reduction in population size leads to decline in genetic diversity by genetic drift and inbreeding. In rare species with small and fragmented populations, the rate may be too low to avoid decreasing population viability due to genetic drift and inbreeding depression (Huenneke, 1991). Populations having low genetic diversity often exhibit an increased rate of extinction (Markert *et al.*, 2010). Sexual reproduction can be limited both by biotic and abiotic factors. It is particularly important to understand the reproductive ecology of rare or endangered plant species to determine whether the cause is intrinsic, i.e. related to the biology of the species such as natural reproductive constraints, or extrinsic to the species, i.e. related to environmental factors. Many species depend on insect pollination as well as seed dispersal by animals, and pollen limitation caused by lack of pollen or pollinators has been shown to be an important factor limiting seed set (Wilcock and Neiland, 2002; Pearse *et al.*, 2015). However, in most cases abiotic factors such as climate and soil conditions directly control over-flowering and fruit set (Pigott, 1992).

3.2 Breeding Systems in Plants

Breeding systems in plants refers to the attributes of the flowers within an individual that may influence transfer of genetic material (gamete) among conspecifics. In other words, breeding systems refers to the diverse ways plants accomplish different stages of reproduction leading to the generation of propagules of the next generation. In angiosperms, flowers are the reproductive structures and represent unique elements of heredity and evolution. Unlike most higher animals, plants are non-motile and cannot choose their mates to share and exchange genetic material. However, plants have evolved diverse mechanisms to control the population genetic structure through their mating patterns. Breeding systems influence many important aspects in a population's ecology and evolution (Charlesworth, 2006). Inbreeding and genetic drift increase in small populations because of fewer numbers of contributing individuals in each generation (Keller and Waller, 2002). The inbreeding process rapidly increases homozygote frequencies, and thus individuals will express recessive or partially recessive deleterious mutations. As a result, species suffer from lower survival, reduced fertility levels or inbreeding depression (Keller and Waller, 2002; Wright *et al.*, 2008). This affects lower mutation frequencies and leads to reduced inbreeding depression, ultimately having a negative impact on the survival and fertility of a species (Byers and Waller, 1999).

Modes of sexuality in plant populations can be hermaphrodite (only hermaphrodite or bisexual individuals), monoecious (only monoecious individuals), dioecious (only dioecious individuals), gynodioecious (both hermaphrodite and gynoecious individuals), androdioecious (both hermaphrodite and androecious individuals) and trioecious (hermaphrodite, pistilate and staminate individuals) (Dellaporta and Calderon-Urrea, 1993). Moreover, sexuality in individual plants can take the form of being gynoecious (plant bearing only pistilate flowers), androecious (plant bearing only staminate flowers), gynomonoecious (plant bearing both hermaphrodite and pistillate flowers), andromonoecious (plant bearing both hermaphrodite and staminate flowers) and trimonoecious or polygamous (hermaphrodite and pistillate and staminate flowers) (Dellaporta and Calderon-Urrea, 1993). Occurrence of different mating systems is not uniformly distributed across angiosperms. It is estimated that only about 6% of flowering plants are dioecious (Renner, 2014) and it remains unclear why dioecy is so rare in angiosperms (Kafer *et al.*, 2017). A recent study showed that bisexual flowers in angiosperms are ancestral, and evolved to unisexual flowers several times and independently (Sauquet *et al.*, 2017).

Androdioecy, the co-occurrence of males and hermaphrodites in a sexually reproducing population is regarded as an exceedingly rare phenomenon (Pannell, 2002a; Weeks, 2012). Although Darwin (1877) reported gynodioecy, i.e. occurrence of females and hermaphrodites, in his book *The Different Forms of Flowers*, the existence of androdioecy in nature remained unnoticed for a long period of time. However, occasional instances of androdioecy were reported (Yampolsky and Yampolsky, 1922). The occurrence of gynodioecy is reported in approximately 10% of angiosperms (Delannay, 1978) whereas cases of androdioecy have so far been reported in very few plant species (Pannell, 2002b). Studies on androdioecy have gained much attention since the appearance of its first confirmed report (Liston *et al.*, 1990) in *Datisca glomerata* (Datiscaceae). Although androdioecy is rare, its maintenance and evolution has broad implications for the advancement of breeding system theory (Charlesworth, 1993). Theory predicts that males will only invade a cosexual population if the fitness of a male is greater than twice that of a cosexual individual (Charlesworth, 1993). Self-fertilization will further reduce the advantage of maleness over cosexuality by reducing the opportunity for males to fertilize cosexual individuals and hence reduce the evolution of androdioecy (Fritsch and Rieseberg, 1992; Pannell, 1997). Another report, however, suggests that many cases of androdioecy (and near-androdioecy) have evolved from dioecy (males and females) rather than hermaphroditism (Wolf *et al.*, 2001). Pannell (2002b), in a review, considered androdioecy to be evolved from dioecy as a result of selection for self-fertile hermaphroditism.

During the last few decades, many plant species have been described as androdioecious on the basis of morphology. However, detailed studies have revealed many such species to be functionally dioecious or cryptic androdioecious with sterile pollen of morphologically hermaphrodite flowers (Mayer and Charlesworth, 1991). On the other hand, very few plant species such as *Datisca glomerata* (Liston *et al.*, 1990), *Mercuria lisannua* (Pannell, 1997) and *Schizopepon bryoniaefolius* (Akimoto *et al.*, 1999) were confirmed to be functionally androdioecious (where both male and hermaphrodite pollens are functional) through pollen fertility and breeding experiments. Another recent study showed clear sexual dimorphism into male and hermaphrodite individuals in *Lloydia oxycarpa* (Liliaceae) in the same population but did not demonstrate significant reproductive advantages of the male flowers relative to the hermaphrodites (Niu *et al.*, 2017). Moreover, it was observed that a considerable number of *Lloydia oxycarpa* individuals changed their sex in successive years and this therefore suggested that the species is not androdioecious but rather a rare case of size-dependent gender diphasy (Niu *et al.*, 2017).

3.3 Breeding Systems in *Gymnocladus assamicus*

Breeding systems in the genus *Gymnocladus* are confusing and described either as polygamous or dioecious, or with unisexual flowers (Herendeen *et al.*, 2003; Dezhao *et al.*, 2010; Watson and Dallwitz, 2018). Flowers in the species *Gymnocladus* and *Gleditsia* are also described either as bisexual or, rarely, as unisexual (Azani *et al.*, 2017). *Gymnocladus chinensis* has been described as polygamous (Jia, 1875). Breeding systems influence many important aspects of population ecology and evolution (Charlesworth, 2006). For this study, extensive initial field trips were made and both fruiting and non-fruiting mature *G. assamicus* trees were documented in the natural population in its native range. It was unclear whether this was due to irregularities in flowering and fruiting behaviour or anthropogenic activities such as logging of branches for firewood. Furthermore, no detailed information was available on floral biology, sexuality and breeding systems of *G. assamicus*. Therefore, an attempt was made to study the reproductive ecology and understand the breeding systems as well as the reproductive strategies of the species. Knowledge on reproductive ecology and associated parameters is directly and indirectly associated with successful fruit and seed production of a species leading to regeneration and long-term survival. A detailed study on reproductive ecology of *G. assamicus* is presented in this chapter, highlighting its sexual dimorphism, breeding system and reproductive strategies. Special emphasis is given to the reproductive characters of the two plant morphs to (a) quantify existing male and hermaphrodite populations; (b) study their reproductive as well as vegetative phenology; (c) examine generalized floral features and stages of floral development; (d) enumerate male and hermaphrodite functions of each sexual phenotype; and (e) finally to study the breeding system.

3.4 Materials and Methods

3.4.1 Study trees

Since *G. assamicus* is a critically endangered tree species with a declining population size, only a limited number of flowering individuals were available for the present study. Mature *G. assamicus* trees are distributed sporadically either as solitary trees or in small populations consisting of one to seven trees. Altogether 28 mature trees were documented from Dirang area in the West Kameng district of Arunachal Pradesh, India. Regular field visits were carried out during the years 2004–2007 to record different phenological events of *G. assamicus* and its associated vegetation. Preliminary observation revealed nine fruiting trees while 19 individuals remained fruitless and were predicted to be males. Spatial distribution of mature male and hermaphrodite individuals in each site is presented in Fig. 3.1.

3.4.2 Reproductive phenology and floral biology

All the 28 trees were included in a floral biology study, while nine individuals from two sites (Dambla village (DM) and Moishing (MS)) were exploited for a reproductive ecology study. However, 10 individuals of Lishpa Village-I and Lishpa Village-II were excluded from the present study since no flowering could be observed during the study period. Phenological events such as budbreak, leaf flushing, shedding of leaves, flowering, fruiting and fruit set were recorded at regular intervals. Daily observations were made throughout the day during the peak flowering period (1–15 April) to determine floral initiation, opening of flower, anthesis, fertilization and fruit set. Subsequent events such as fruit development and fruit maturation were recorded on a weekly basis. Number of flowers per inflorescence (n=20 per tree) was counted for each tree from randomly selected branches. Morphology of flowers and their parts was studied using a hand lens. Flower longevity was determined from 20 flowers per tree by following each developmental stage until wilting. Growth and development of different floral parts were also recorded twice daily (at around 0600 h and 1800 h). Nectar volume (μl) of both the floral morphs was measured by extracting the nectar by graduated micropipettes at different growth stages (n=20). Nectar secretion patterns in male and hermaphrodite flowers were determined using 20 randomly selected flowers from 20 bagged inflorescences. Measurements were made at 0600 h and 1800 h every day during the entire lifetime of the flower. Floral morphometric measurements were taken using a hand-held digital caliper (Mitutoyo, Japan).

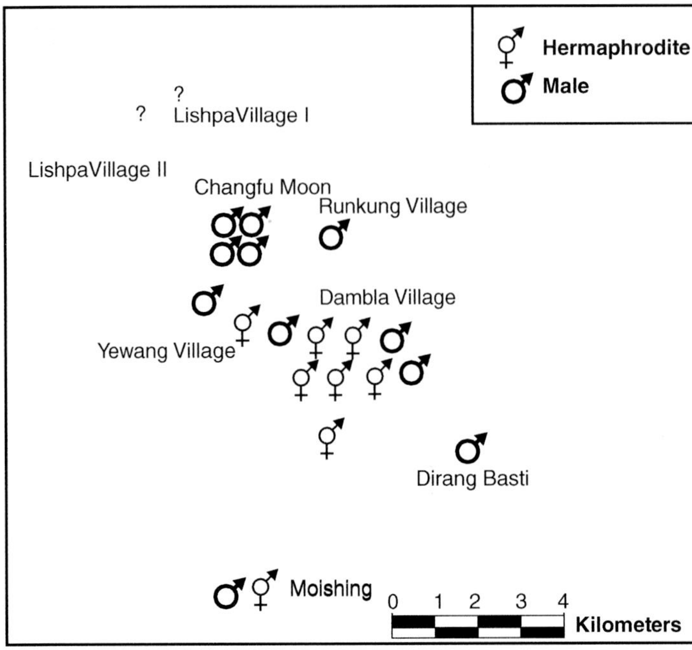

Fig. 3.1. Spatial distribution of male and hermaphrodite trees. (1) Moishing; (2) Dirang Village; (3) Changfu Moon; (4) Yewang Village; (5) Dambla Village; (6) Jyotinagar; (7) Runkung Village; (8) Lishpa Village I; (9) Lishpa Village II.

3.4.3 Flower visitors

Flower visitors such as insects, birds and mammals were recorded at the Moishing and Dambla sites in the selected trees of both the sexes during peak flowering period (April). Observations were made throughout the day using binoculars. Prominent insect pollinators were trapped with sweep nets and were classified into pollinator groups such as bees, beetles, moths, butterflies, etc. Among insects, only the honey bees were identified up to species level. Insect species were identified by consulting manuals and with help taken from the Zoological Survey of India (ZSI) Itanagar, Arunachal Pradesh, India. Bird species were identified with the help of photographs and relevant field observations and after consultation with a renowned ornithologist of the region (Dr Anwaruddin Choudhury, Rhino Foundation, Guwahati, Assam, India).

For the convenience of study, the floral ontogeny was divided into six stages as illustrated in Fig. 3.2. The pre-anthesal bud stage (dark-violet perianth and closed tepal ready to open) was designated as stage 1 (Table 3.1). During this stage, all the reproductive parts of both the sexes are at pre-anthesis period (time: 0 h). During stage 2, anthesis initiated and tepals opened (time: 24 h) while at stage 3, flowers are

Fig. 3.2. Different developmental stages (S1–S6) of *G. assamicus* flower used for pollination experiment (photo: Baharul I. Choudhury).

almost opened with ca. 15.75 mm long and spreading tepals (time: 48 h). Stage 4 is designated as the fully opened stage in which anther lobes were at the same length as the stigma. At this stage pollen grains were highly fertile (58–60%) along with a highly receptive stigma in the case of bisexual flowers (time: 72 h). At stage 5, flowers begin to wilt, tepals start to curl and reproductive parts enter the drying phase (time: 96 h). Stage 6 is the onset period of floral dying when the perianth tube starts falling, leaving only the pistil (in the case of a bisexual flower). During this stage, pollen fertility ceases and the stigma dries up.

3.4.4 Pollen viability and stigma receptivity test

Pollen viability is essential for successful fertilization and eventually fruit set. There are various methods to assess pollen viability in field conditions. We used a simple but efficient sitting drop culture method (Shivanna and Rangaswamy,

Table 3.1. Stages of *G. assamicus* floral development.

		Dimension of flower			Pollen fertility (%)		Nectar volume (µl)	
Stages	Opening order	Length (mm)	Width (mm)	Stigma receptivity	Male	Hermaphrodite	Male	Hermaphrodite
S1		8.15	2.45	X	X	X	X	X
S2		9.44	2.74	X	X	X	X	X
S3		15.75	3.12	Modest	8–10	10–12	2.5–3	2–2.5
S4		16.52	3.74	Maximum	58–60	58–62	16–18	14–15
S5		16.08	3.86	Weak	6–8	8–10	10–12	9–10
S6		15.98	3.33	X	X	X	X	X

S1–S6 = Stage 1 – Stage 6; X = absent.

1992) to estimate the percentage of pollen germination in the field. Pollen was collected from 10 randomly harvested flowers belonging to different individuals of male and hermaphrodite plants at different developmental stages. Pollen sacs were gently crushed on a microscope slide and individual pollens were germinated in Brewbaker and Kwack's medium (Brewbaker and Kwack, 1963). Very briefly, Brewbaker and Kwack's medium is composed of 10% sucrose, 100 $mg\,l^{-1}$ boric acid, 300 $mg\,l^{-1}$ calcium nitrate, 200 $mg\,l^{-1}$magnesium sulphate and 100 $mg\,l^{-1}$ potassium nitrate. Germinated and ungerminated pollen grains were scored under a compound microscope and the frequencies calculated accordingly. Five replicates were taken both from hermaphrodite and male trees for the pollen viability test. Pollen was considered germinated when there was clear and visible pollen tube growth as seen under a microscope.

Stigma receptivity is equally important for successful fertilization by viable pollen grains. Effective pollination period is directly associated with the duration of stigmatic receptivity (Sanzol and Herrero, 2001). The stigma receptivity in *G. assamicus* was tested by putting a few drops of H_2O_2 on a wet stigma surface following the method of Dafni (1992). Emission of bubbles from the stigma surface with high speed was considered to have strong receptivity while slow emission of bubbles was considered to have weak receptivity (Dafni, 1992).

3.4.5 Pollen morphology

Pollen morphology and its association with pollination mechanisms is well studied (Tanaka *et al.*, 2004; Wang *et al.*, 2009) and has several implications in the seed and fruit production of a species. We investigated pollen morphology of *G. assamicus* using scanning electron microscopy (SEM) to demonstrate morphological variations, if any, between the male and hermaphrodite morphs. Dried and powdered pollen samples were mounted on metal stubs and were gold-coated. The pollen samples of each floral morph were then observed under a scanning electron microscope (LEO 1430VP, Karlzeiz, German) and digital photographs were taken under different magnifications.

3.4.6 Breeding systems

To understand the breeding systems, different pollination treatments were carried out following Dafni (2007). Brief descriptions of the treatments are outlined below.

1. **Open pollination** – fruit set under natural conditions in non-manipulated flowers.
2. **Spontaneous self-pollination** – fruit set in bagged, unopened flowers. For this, randomly selected flowers were bagged to exclude visitors.
3. **Control pollination** – freshly opened flowers were emasculated using fine forceps and re-bagged for artificial pollination. Flowers of other stages borne on the inflorescence were carefully removed before bagging. Manual autogamous, geitonogamous and xenogamous pollinations were carried out in situ during the stigma receptivity period (stage 4; Table 3.1). Manual autogamous and xenogamous pollination were given the main emphasis to verify the phenomenon of androdioecy.
4. **Agamospermic and parthenocarpic fruit set** was also studied by bagging emasculated hermaphrodite flowers without pollination.

Statistical analysis was done following Zar (1974) and using software packages (MS Excel, SYSTAT, STATISTICA and ORIGIN).

3.5 Results

3.5.1 Phenology

G. assamicus trees remain completely leafless for over two months during the winter season (January–February). Leaf flushing takes place in early March and continues for about two weeks. Mature leaves turn bright yellow in September and are visually attractive. Phenology of vegetative phases is similar in both male and hermaphrodite plants. Flowering begins immediately after leaf flushing and lasts for 15–20 days. Flowers in male trees bloom between the last week of March and the second week of April, whereas the peak bloom of hermaphrodite trees takes place in April (Fig. 3.3). Fully mature pods persist on trees until the next flowering season when fruits become ready to disperse by gravity (Fig. 3.4).

3.5.2 Floral sexuality

Both male and bisexual flowers are small (3.43–3.74 mm x 5.50–9.20 mm), tubular, violet-coloured and odourless. Flowers open in the early morning and last

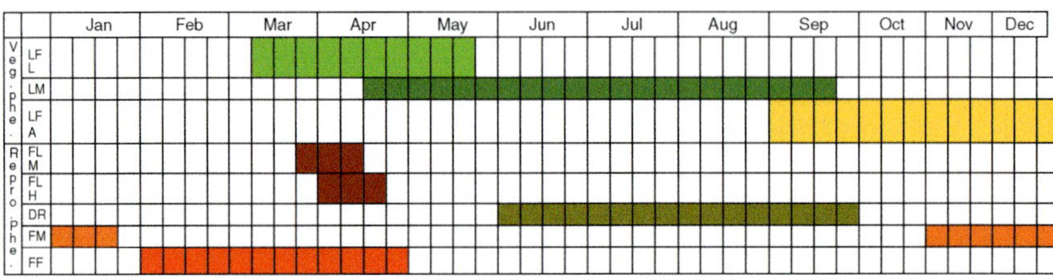

Fig. 3.3. Vegetative and reproductive phenology of male and hermaphrodite *G. assamicus* trees. LFL = Leaf flushing; LM= Leaf maturation; LFA = Leaf fall; FLM = Flowering (male); FLH = Flowering (hermaphrodite); DR = Drupe initiation; FM = Fruit maturation; FF = Fruit fall.

Fig. 3.4. Persistent mature pods (from the previous year) during the next flowering season (photo: Baharul I. Choudhury).

for around 96 hours. Male inflorescence is much larger in size compared to the hermaphrodite inflorescence and bears around three times more flowers than its hermaphrodite counterparts. Average number of flowers per male inflorescence was recorded to be 72.50 (±4.01, n=20) while it was 23.40 (±4.01, n=20) for hermaphrodite flowers.

Pedicellate male flowers are cylindrical, 5.50 mm long and 3.43 mm wide at anthesis. Male flowers show a rudimentary carpel which is consistently present in every flower (Fig. 3.5). Flowers open between 09.00–11.00 h and anthers dehisce after 20.00–24.00 h (Table 3.1). Pollen grains from freshly opened male flowers showed 8–10% germination while the germination rate was 58–60% at stage 4 (Table 3.1). Pollen grain germination percentage then steadily decreased with age and was completely lost at stage 6 (Table 3.1).

Hermaphrodite flowers have longer pedicellate compared to the male flowers and are approximately 9.20 mm long and 3.74 mm wide at anthesis. During floral development, the filaments as well as the pistil enlarge simultaneously and at fully matured stage, longer anther lobes reach above the receptive surface of the stigma, while shorter ones remain at the level of the stigma (Fig. 3.6). The pistil in

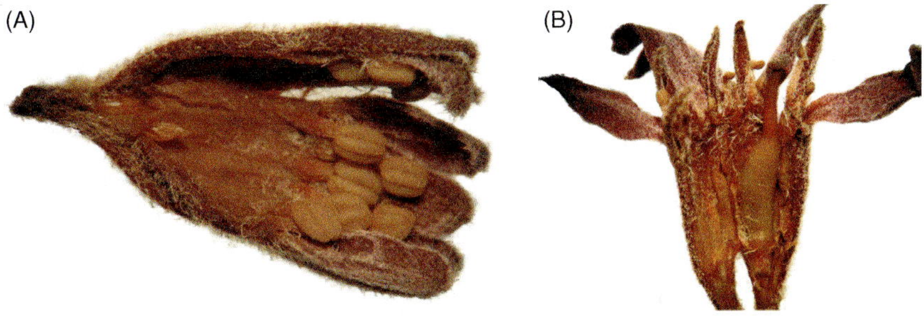

Fig. 3.5. Flowers showing only anther lobes and rudimentary carpel in male (A) and both carpel and anther lobes in hermaphrodite (B) flowers.

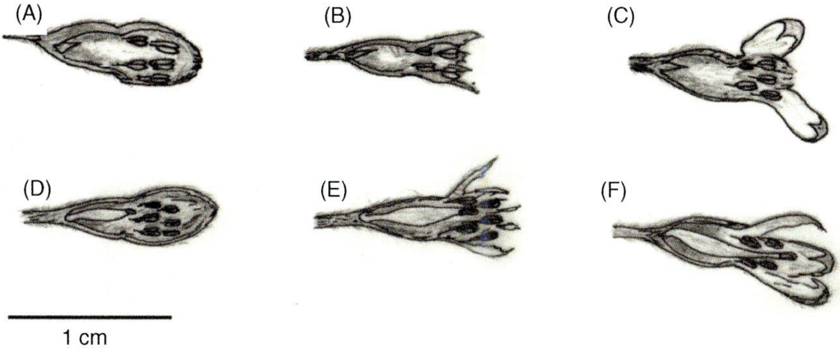

Fig. 3.6. Three developmental stages of male (A, B, C) and hermaphrodite (D, E, F) flowers.

hermaphrodite flowers is 9–10 mm long at anthesis, slightly arched, thick, compressed and well demarcated into stigma, style and ovary. The style is straight, slightly thick and compressed with oblique stigma. The stigma surface is slightly slanted, papillate, with wet sticky exudates. Stigma receptivity was found highest at stage 4 (Table 3.1). Unilocular superior ovary contains 2–9 ovules. Detailed flower morphometric characteristics are presented in Table 3.2.

3.5.3 Pollen fertility

Pollen of both the morphs germinated well in Brewbaker and Kwack's medium (Fig. 3.7). Pollen viability percentage of hermaphrodite flowers was similar to that of their male counterparts. Flowers open between 14.00–16.00 h. Maximum pollen germination percentage (58–62%) was observed at stage 4. The trend of loss of pollen germination percentage in hermaphrodite flowers was also similar to that of male flowers. Pollen germination percentages at different floral stages of both the floral morphs as scored under a microscope are presented in Table 3.1.

3.5.4 Nectar production

Flowers start nectar production when they attain the developmental stage 2 (Table 3.1). Though the pattern of nectar secretion is almost uniform among different flowers, nectar volume varied significantly ($p<0.05$) between the two sexes of the flower (Table 3.1). Flowers of both the morphs produced the highest quantity of nectar (male 16–18 μl; hermaphrodite 14–15 μl) at stage 4.

3.5.5 Scanning election microscopy of pollen

Scanning electron microscopic observation of pollen grains of both the morphs revealed that pollens are of polycolpate type and are psilate. Pollen grain from male flowers (Fig. 3.8A) and hermaphrodite flowers (Fig. 3.8B) is more or less similar morphologically.

3.5.6 Breeding system

The results of different breeding experiments carried out to assess the breeding system of *G. assamicus* are as follows.

Open pollination: fruit set percentage for spontaneous open pollination (un-bagged and non-manipulated flowers) was 48.31 (±2.75). On the other hand, manual autogamous pollination resulted in 74.91% (±2.92) fruit set, whereas it was 65.59% (±2.82) fruit set in geitonogamous pollination. There was no significant fruit set difference between the two study sites (MS and DM) in all the above experiments.

Table 3.2. Floral characteristics of *G. assamicus*.

Features	Inflorescence Length (cm)	Inflorescence Breadth (cm)	No. of flower per inflorescence	Average flower production	Biomass at fully opened stage (gm)	Length of corolla tube (mm)	Width of corolla tube (mm)	Length of pistil (mm)	Length of anther lobe (mm)	
									Longer	Shorter
Male	12 (±0.81)	2.92 (±0.14)	72.50 (±4.01)	238227.92	0.049 (±0.002)	5.50 (±0.027)	3.43 (±0.038)	N/A	8.99 (±0.065)	8.80 (±0.015)
Hermaphrodite	3.38 (±0.11)	2.26 (±0.06)	23.40 (±1.41)	80521.98	0.088 (±0.004)	9.20 (±0.017)	3.74 (±0.014)	10.84 (±0.054)	9.67 (±0.050)	9.07 (±0.025)

±S.E. n=20.

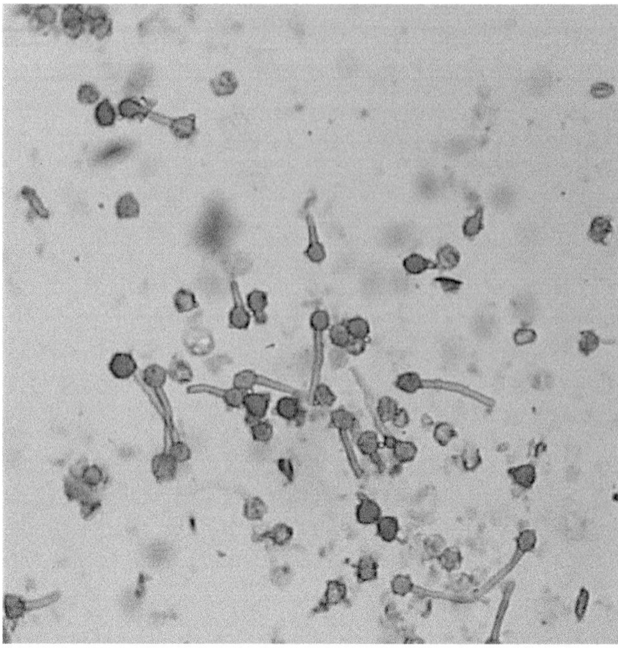

Fig. 3.7. Pollen germination in Brewbaker and Kwack's medium.

Xenogamy: xenogamous pollination was carried out by taking pollen from male as well as hermaphrodite individuals. Fifty inflorescences (25 in each site) were randomly considered for each xenogamous pollination with crosses between male and hermaphrodite (Fig. 3.9A) and hermaphrodite and hermaphrodite (Fig. 3.9B) and their results were observed. Fruit set in crosses between male and hermaphrodite flowers showed 57.50% (±3.85), whereas fruit set in the latter was 68.40% (±3.53). Details of different breeding experiments, fruit set percentage and differences of fruit set with open pollination are presented in Table 3.3.

3.5.7 Pollinator spectra

Altogether nine species of anthophilous insects, which include the social bees, beetles and butterflies, were encountered frequently from flowering *G. assamicus* trees. Social bees were composed of only two species, *Apis cerana* and *Apis dorsata*. *Apis dorsata* was more commonly found in both the sites followed by *Apis cerana*. Three species of leaf cutter bees (*Megachile* spp.) were also recorded from different flowering *G. assamicus* trees. One species of flower chaffer and two species of butterfly also visited the tree species. Among vertebrates, only one nectarivorous bird the 'green-tailed sunbird' or 'Nepal yellow-backed sunbird' (*Aethopyga nipalensis*) was found to be very active during peak flowering periods (Fig. 3.10). Table 3.4 shows a list of different flower visitors.

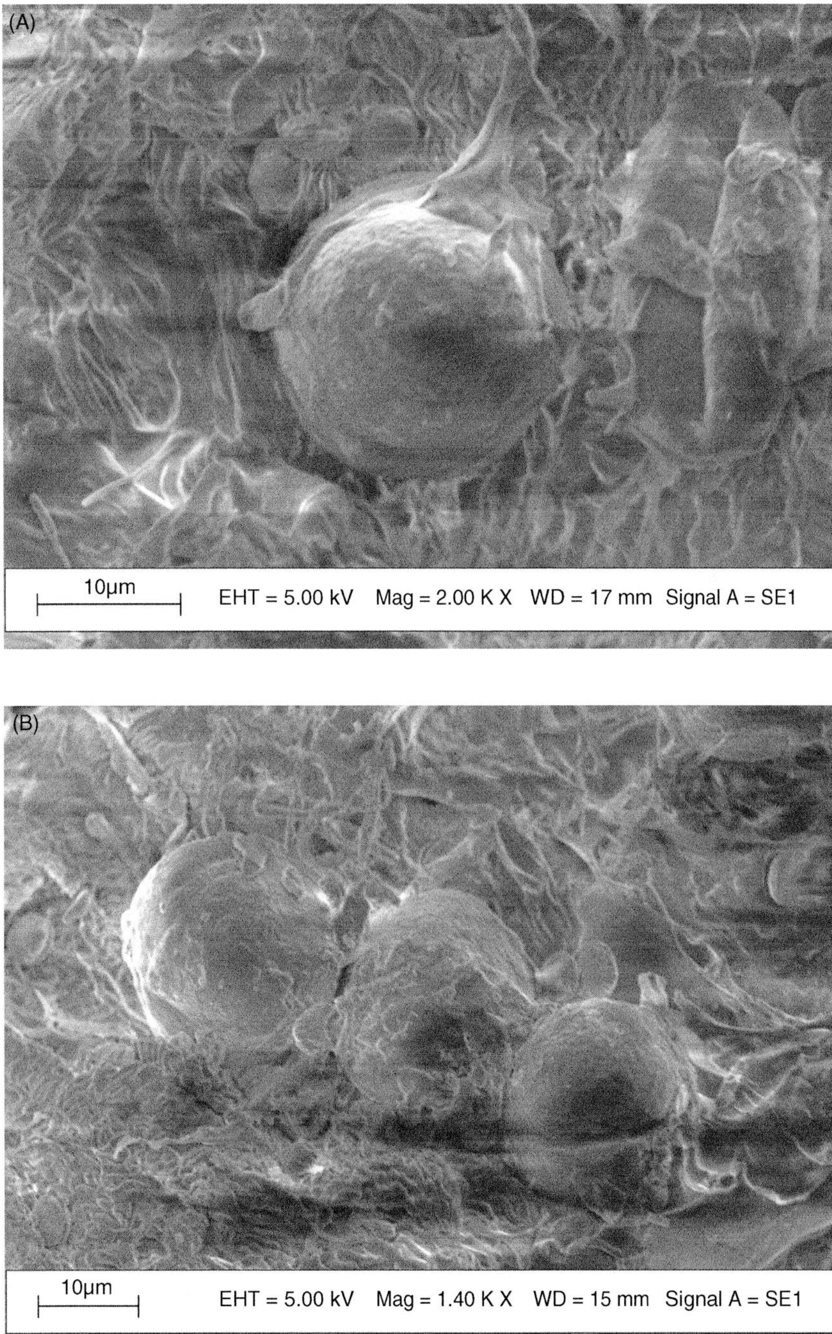

Fig. 3.8. Scanning Electron Micrograph of pollen. (A) Polar view of male pollen showing polycolpate structure; (B) Distal view of hermaphrodite pollen.

Fig. 3.9. Fruit set in xenogamous pollination. (A) Pollen from male flower; (B) Pollen from hermaphrodite flower.

Table 3.3. Results of controlled pollination experiments.

Treatment	Number of flowers pollinated	Number of flowers producing fruit	Average fruit set percentage (±SE)	*Probability level of fruit set
Open pollination	302	146	48.31 (±2.75)	
Spontaneous self-pollination	396	157	36.64 (±2.48)	
Hand pollination	143	107	74.91 (±2.92)	F=10.325,
Geitonogamy	160	105	65.59 (±2.82)	*p*<0.001
Xenogamy with male	134	78	57.50 (±3.85)	
Xenogamy with hermaphrodite	147	99	68.40 (±3.53)	

* Differences in percentage fruit set among treatments were highly significant.

Although a total of 28 mature *G. assamicus* trees were located from the study area, only nine hermaphrodites and nine male individuals were usable for the study. In two populations (Moishing and Yewang village) one male and one bisexual tree were available, and the ratio was observed to be 1:1, while in four sites (Dambla village, Runkung village, Lishpa village-I and Lishpa village-II) only bisexual individuals were present. Two male and five hermaphrodite individuals of *G. assamicus* were in the Changfu Moon site. The frequency of male and hermaphrodite in majority of the populations were deviated from a 1:1 ratio. On the other hand, it was found that single hermaphrodite trees in multiple sites successfully set fruit and seeds, which indicates that hermaphrodite trees are self-compatible for sexual reproduction and seed set was successful even in the absence of surrounding male trees. Moreover, frequency of the male tree is lower (32.14%) than the hermaphrodites over all the sites. Although the present study is on a critically endangered species with isolated trees that cannot be grouped into 'populations', the pattern of overall sex ratio found was similar to that of other functionally androdioecious species such as

Fig. 3.10. Male green-tailed sunbird (*Aethopyga nipalensis*) visiting *G. assamicus* trees during flowering season (A, B); (C) A flower pierced by a green-tailed sunbird (photo: Baharul I. Choudhury).

Table 3.4. Flower visitors of *G. assamicus*.

Sl. No.	Common name	Order	Family	Scientific name
1	Honey bee	Hymenoptera	Apidae	*Apis cerana* Fabr.
2	Giant honey bee	Hymenoptera	Apidae	*Apis dorsata* Fabr.
3	Bumblebee	Hymenoptera	Apidae	*Bombus* sp.
4	Leaf cutter bee	Hymenoptera	Megachilidae	*Megachile* sp.
5	Leaf cutter bee	Hymenoptera	Megachilidae	*Megachile* sp.
6	Leaf cutter bee	Hymenoptera	Megachilidae	*Megachile* sp.
7	Flower chaffer	Coleoptera	Scarabaeidae	Unidentified
8	Moth/butterfly	Lepidoptera	Unidentified	Unidentified
9	Moth/butterfly	Lepidoptera	Unidentified	Unidentified
10	Green-tailed sunbird	Falconiformes	Accipitridae	*Aethopyga nipalensis*

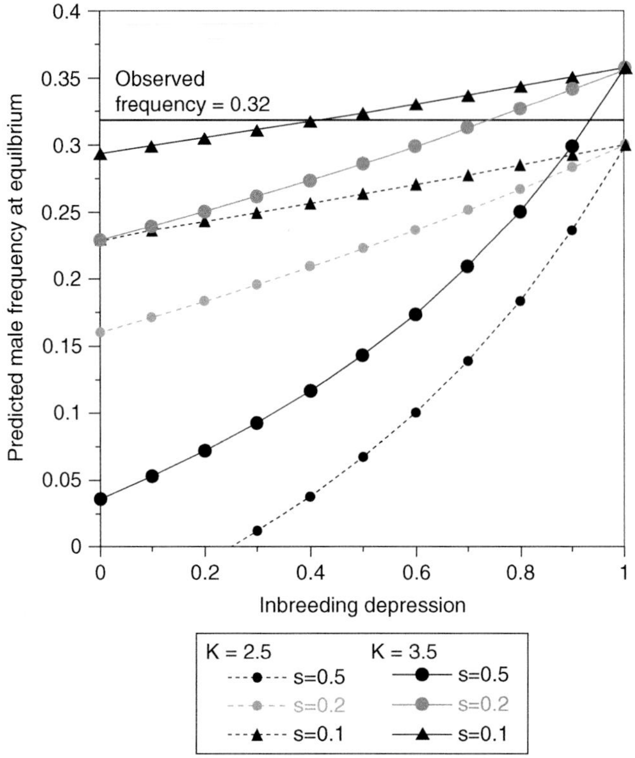

Fig. 3.11. Predicted male frequency versus inbreeding depression. K values slightly below (K=2.5) or above (K=3.5) the estimated K value (K=2.9) based on the observed number of flowers in male and female trees used. Three selfing rates (s=0.1, 0.2 and 0.5) were chosen for this prediction. The observed frequency of male trees in the population was 32.14%.

Datisca glomerata (<25%) (Liston *et al.*, 1990), *Mercurialis annua* (<30%) (Pannell, 1997) and *Schizopepon bryoniaefolius* (up to 28.3%) (Akimoto *et al.*, 1999).

3.5.8 Equilibrium male frequency in the population

The observed male frequency was slightly higher than the expected equilibrium frequency of males under theoretical expectations with high levels of inbreeding depression (Fig. 3.11).

3.6 Discussion

Phenological studies on the vegetative phase of *G. assamicus* in both the sexes showed similar patterns, while reproductive phenology showed slightly different but negligible variation. However, overlapping of flowering phenology between the two sexes was observed during this study. Such overlapping thus provides an opportunity for pollen transfer from male to hermophrodite trees thereby allowing xenogamous pollination in such reproductive system. Flower count revealed that male individuals produced as many as 238,227 flowers per tree which is about three times per hermaphrodite tree (about 80,521) (Table 3.2). This may act as an advantage in pollen transfer from male trees, avoiding continuous selfing and inbreeding depression in those sites where male and hermaphrodite individuals coexist

Floral morphometric characteristics showed that male flowers are smaller in size than hermaphrodite flowers. However, developmental stages, nectar secretion pattern and floral life span were similar in both the morphs (Table 3.1). Pollen germination percentage showed a gradual increase from around 10% at stage 3 up to a maximum of 60% at stage 4, which then reduced to zero at stage 6 (Table 3.1). This observation revealed that both the sexes of *G. assamicus* have viable pollen grains. The largest nectar volume in both the morphs was recorded at stage 4 (ca. 16μl/flower) and at this stage the stigma was found to be in the highest receptive state (Table 3.1).

Different breeding experiments revealed that hermaphrodite individuals of *G. assamicus* are self-compatible and autogamous and they can set fruit up to 76.81%. Moreover, four out of nine sites had only hermaphrodite trees and were producing fruit and seed consistently. Xenogamous crosses with male pollen sources, on the other hand, also showed 56.85% fruit set, thereby confirming fertility of the male pollen grains. The results of different breeding experiments (Fig. 3.12), provided firm evidence that *G. assamicus* is a functionally androdioecious species.

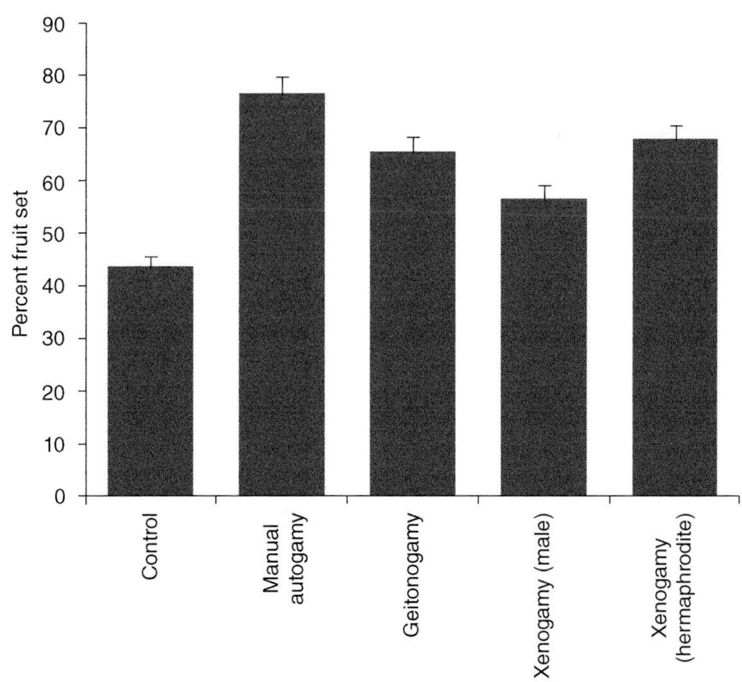

Fig. 3.12. Fruit set percentage of different pollination experiments.

There are a handful of species displaying androdioecious mating systems. Theoretical models explain the rarity of androdioecy by predicting that males must sire at least twice as many successful offspring as hermaphrodites (Charlesworth and Charlesworth, 1978; Charlesworth 1984). Hermaphroditism is thought to be the ancestral breeding system of angiosperms (Lewis, 1942; Bawa, 1980; Charlesworth, 1985) and several models that explain the evolutionary basis of dioecism have been postulated (Bawa and Beach, 1981). Pannell (2002b) suggested that androdioecy in most of the reported species most likely evolved from dioecy through the spread of pollen-producing females rather than from a hermaphroditic ancestor through the spread of female-sterility mutations. However, based on our observations and empirical data such as rudimentary carpel in males, almost similar percentages of pollen viability in both the sexes and more or less equal fruit set from both the xenogamous pollinations, it may be anticipated that androdioecy in *G. assamicus* is derived from hermaphroditism and not from dioecy. Such cases have been reported in the olive family (Oleaceae) in which androdioecy appears to have evolved from hermaphroditism multiple times through the spread of female-sterile mutations (Wallander, 2001). Similar hypotheses were also proposed by Lepart and Dommee (1992) while studying *Phillyrea anguistifolia* in which hermaphrodites with perfect flowers coexist with female-sterile males. In the case of *Cardamine mara* (Brassicaceae), a combination of sexual reproduction and increased asexual proliferation by female-sterile individuals was observed which probably explains the invasion and maintenance of female sterility in otherwise hermaphroditic populations (Tedder *et al.*, 2015). Nectar production in both males and hermaphrodites showed almost similar patterns except the higher volume in males. This may be a significant contribution made by the male tree to attract more pollinators for its effective pollen transfer to the hermaphrodites. A similar phenomenon is also evidenced in the case of a protandrous herb (*Chrysothemis friedrichsthaiana*, Gesneriaceae) (Carlson, 2007).

Present observations on breeding systems of *G. assamicus* revealed that hermaphrodite trees are self-compatible for production of fruits and viable seeds. Moreover, pollen grains of male trees are also viable and can contribute to fruit and seed production. On the basis of our observations, empirical data, and according to the definition of functional androdioecy by Pannell (2002a), *G. assamicus* is confirmed to bear clear sexual dimorphism, with males lacking female function and hermaphrodites with both sexual functions. Therefore, it is confirmed that *G. assamicus* is a functional androdioecious species (Choudhury *et al.*, 2014).

3.7 Evolution of *G. assamicus* Breeding System

Dioecy is a rare sexual system and has been suggested as an evolutionary dead end (Bull and Charnov, 1985; Heilbuth, 2000); however, recent studies demonstrate that transitions from dioecy to other breeding systems are possible (Barrett, 2013; Goldberg *et al.*, 2017; Zhang *et al.*, 2018). Because of the rarity of androdioecy, it is predicted to be a short-lived and intermediate breeding system towards the evolution of dioecy (Lloyd, 1975; Bawa, 1980; Charlesworth, 1984; Wolf and Takebayashi, 2004; Astrop *et al.*, 2012) or may have evolved from dioecy as a result of selection for male function in females for reproductive assurance during colonization (Rieseberg *et al.*, 1992; Pannell, 1997; Akimoto *et al.*, 1999; Wolf *et al.*, 2001). More interestingly, most examples of androdioecy appear to have evolved from dioecy through the spread of pollen-producing females rather than from a hermaphroditic ancestor through the spread of female-sterility mutations (Pannell, 2002b).

Male *G. assamicus* individuals produce more than three times as many flowers as hermaphrodites. A significantly higher rate of flower production, and in turn pollen production, by male individuals may benefit pollination success, avoid inbreeding and be beneficial for outcrossing success. Similar results were also observed in other androdioecious species such as *D. glomerata* and the phenomenon has been shown to exhibit outcrossing advantage (Fritsch and Rieseberg, 1992; Rieseberg *et al.*, 1993). Studies demonstrate that the most probable route of androdioecy evolution is from dioecy through the evolution of a male function in females; however, the role of female sterility and subsequent invasion into hermaphrodite populations has also been shown in several plant species (Wallander, 2001; Tedder *et al.*, 2015). In *G. assamicus*, hand pollination increases the fruit set suggesting pollination limitation for fruit set. Phylogenetic studies in conjunction with mating system analyses of related species paternity tests will help to resolve the question if androdioecy in *G. assamicus* evolved from hermaphrodites or dioecious progenitors.

3.8 References

Akimoto, J., Fukuhara, T. and Kikuzawa, K., 1999. Sex ratios and genetic variation in a functionally androdioecious species, *Schizopepon bryoniaefolius* (Cucurbitaceae). *American Journal of Botany* 86(6): 880–886.

Astrop, T.I., Park, L.E., Brown, B. and Weeks, S.C., 2012. Sexual discrimination at work: spinicaudatan 'clam shrimp' (Crustacea: Branchiopoda) as a model organism for the study of sexual system evolution. *Palaeontologia Electronica* 15(2): 1–15.

Azani, N., Babineau, M., Bailey, C.D., Banks, H., Barbosa, A.R., Pinto, R.B., Boatwright, J.S., Borges, L.M., Brown, G.K., Bruneau, A. and Candido, E., 2017. A new subfamily classification of the Leguminosae based on a taxonomically comprehensive phylogeny. The Legume Phylogeny Working Group (LPWG). *Taxon* 66(1): 44–77.

Baack, E., Melo, M.C., Rieseberg, L.H. and Ortiz-Barrientos, D., 2015. The origins of reproductive isolation in plants. *New Phytologist* 207(4): 968–984.

Bailey, P. and Kevan, P.G., 2017. Reproductive strategies matter for rare plant conservation: pollination, phenology, and mating in an endemic plant of the Sky Islands of Arizona, *Erigeron lemmonii* A. Gray (Lemmon's Fleabane)(Asteraceae). *Western North American Naturalist* 77(3): 385–397.

Baker, H.G., 1959. Reproductive methods as factors in speciation in flowering plants. *Cold Spring Harbor Symposia on Quantitative Biology* 24: 177–191.

Barrett, S.C., 2013. The evolution of plant reproductive systems: how often are transitions irreversible? *Proceedings of the Royal Society B: Biological Sciences*, 280(1765).

Bawa, K.S., 1980. Evolution of dioecy in flowering plants. *Annual Review of Ecology and Systematics* 11(1): 15–39.

Bawa, K.S. and Beach, J.H., 1981. Evolution of sexual systems in flowering plants. *Annals of the Missouri Botanical Garden.* 68(2): 254–274.

Brewbaker, J.L. and Kwack, B.H., 1963. The essential role of calcium ion in pollen germination and pollen tube growth. *American Journal of Botany* 50(9): 859–865.

Bull, J.J. and Charnov, E.L., 1985. On irreversible evolution. *Evolution* 39(5): 1149–1155.

Byers, D.L. and Waller, D.M., 1999. Do plant populations purge their genetic load? Effects of population size and mating history on inbreeding depression. *Annual Review of Ecology and Systematics* 30(1): 479–513.

Carlsen, T.M., Espeland, E.K. and Pavlik, B.M., 2002. Reproductive ecology and the persistence of an endangered plant. *Biodiversity & Conservation* 11(7): 1247–1268.

Carlson, J.E., 2007. Male-biased nectar production in a protandrous herb matches predictions of sexual selection theory in plants. *American Journal of Botany* 94(4): 674–682.

Charlesworth, B. and Charlesworth, D., 1978. A model for the evolution of dioecy and gynodioecy. *The American Naturalist* 112(988): 75–997.

Charlesworth, D., 1984. Androdioecy and the evolution of dioecy. *Biological Journal of the Linnean Society*, 22(4): 333–348.

Charlesworth, D., 1985. Distribution of dioecy and self-incompatibility in angiosperms. In J. Greenwood and M. Slatkin (eds) *Evolution: Essays in Honour of John Maynard Smith*, pp. 237–268.

Charlesworth D., 1993. Save the males. *Current Biology* 3: 155–157.

Charlesworth, D., 2006. Evolution of plant breeding systems. *Current Biology* 16(17): R726–R735.

Choudhury, B.I., Khan, M.L., Arunachalam, A. and Das, A.K., 2007. Ecology and conservation of the critically endangered tree species *Gymnocladus assamicus* in Arunachal Pradesh, India. *International Journal of Ecology*, http://dx.doi.org/10.1155/2007/59282, accessed 7 June 2019.

Choudhury, B.I., Khan, M.L. and Dayanandan, S., 2014. Functional androdioecy in critically endangered *Gymnocladus assamicus* (Leguminosae) in the eastern Himalayan region of northeast India. *PloS one* 9(2): e87287.

Crow, J.F., 1992. An advantage of sexual reproduction in a rapidly changing environment. *Journal of Heredity* 83(3): 169–173.

Dafni, A., 1992. *Pollination Ecology: A Practical Approach.* Oxford University Press, Oxford.

Dafni, A., 2007. *Manuals for the Field Workshop on Pollination Ecology.* Ashoka Trust for Research in Ecology and the Environment (ATREE), Bangalore, India.

Darwin, C., 1877. *The Different Forms of Flowers on Plants of the Same Species.* John Murray London.

Delannay, X., 1978. La gynodioeci chez les angiospermes. *Naturalistes Belges* 99: 169–181.

Dellaporta, S.L. and Calderon-Urrea, A., 1993. Sex determination in flowering plants. *The Plant Cell* 5(10): 1241–1251.

Dezhao, C., Dianxiang, Z. and Larsen, K., 2010. GYMNOCLADUS Lamarck, Encycl. 1: 733. 1785, nom. cons. *Flora of China* 10: 36.

Forrest, J.R., 2015. Plant–pollinator interactions and phenological change: what can we learn about climate impacts from experiments and observations? *Oikos* 124(1): 4–13.

Frankham, R., Ballou, J.D. and Briscoe, D.A., 2010. *Introduction to Conservation Genetics* (2nd edn). Press Syndicate for the University of Cambridge, Cambridge.

Fritsch, P. and Rieseberg, L.H., 1992. High outcrossing rates maintain male and hermaphrodite individuals in populations of the flowering plant *Datisca glomerata. Nature* 359(6396): 633.

Goldberg, E.E., Otto, S.P., Vamosi, J.C., Mayrose, I., Sabath, N., Ming, R. and Ashman, T.L., 2017. Macroevolutionary synthesis of flowering plant sexual systems. *Evolution* 71(4): 898–912.

Hegland, S.J., Nielsen, A., Lázaro, A., Bjerknes, A.L. and Totland, Ø., 2009. How does climate warming affect plant-pollinator interactions? *Ecology Letters* 12(2): 184–195.

Heilbuth, J.C., 2000. Lower species richness in dioecious clades. *The American Naturalist* 156(3): 221–241.

Herendeen, P.S., Lewis, G.P. and Bruneau, A., 2003. Floral morphology in caesalpinioid legumes: testing the monophyly of the 'Umtiza clade'. *International Journal of Plant Sciences* 164(S5–S393–S407).

Huenneke, L.F., 1991. Ecological implications of genetic variation in plant populations. *Genetics and Conservation of Rare Plants* 31: 31–2.

Jia, F.Z., 1875. *Gymnocladus chinensis* Baillon. *Bulletin Mensuel de la Societe Linneenne de Paris* 1(34).

Kafer, J., Marais, G.A. and Pannell, J.R., 2017. On the rarity of dioecy in flowering plants. *Molecular Ecology* 26(5): 1225–1241.

Keller, L.F. and Waller, D.M., 2002. Inbreeding effects in wild populations. *Trends in Ecology & Evolution* 17(5): 230–241.

Kua, C.S. and Cannon, C.H., 2017. Patterns of genomic diversification reflect differences in life history and reproductive biology between figs (Ficus) and the stone oaks (Lithocarpus). *Genome* 60(9): 756–761.

Lepart, J. and Dommee, B., 1992. Is *Phillyrea angustifolia* L.(Oleaceae) an androdioecious species? *Botanical Journal of the Linnean Society* 108(4): 375–387.

Lewis, D., 1942. The evolution of sex in flowering plants. *Biological Reviews* 17(1):46–67.

Liston, A., Rieseberg, L.H. and Elias, T.S., 1990. Functional androdioecy in the flowering plant *Datisca glomerata*. *Nature* 343(6259): 641–642.

Lloyd, D.G., 1975. The maintenance of gynodioecy and androdioecy in angiosperms. *Genetica* 45(3): 325–339.

Markert, J.A., Champlin, D.M., Gutjahr-Gobell, R., Grear, J.S., Kuhn, A., McGreevy, T.J., Roth, A., Bagley, M.J. and Nacci, D.E., 2010. Population genetic diversity and fitness in multiple environments. *BMC Evolutionary Biology* 10(1): 205.

Mayer, S.S., and Charlesworth, D., 1991. Cryptic dioecy in flowering plants. *Trends in Ecology & Evolution* 6.10: 320–325.

Moza, M.K. and Bhatnagar, A.K., 2007. Plant reproductive biology studies crucial for conservation. *Current Science* 92(9): 1207.

Niu, Y., Gong, Q., Peng, D., Sun, H. and Li, Z., 2017. Function of male and hermaphroditic flowers and size-dependent gender diphasy of *Lloydia oxycarpa* (Liliaceae) from Hengduan Mountains. *Plant Diversity* 39(4): 187–193.

Pannell, J., 1997. Variation in sex ratios and sex allocation in androdioecious Mercurialis annua. *Journal of Ecology*: 57–69.

Pannell, J.R. 2002a. What is functional androdioecy? *Functional Ecology* 16(6): 862–865.

Pannell, J.R. 2002b. The evolution and maintenance of androdioecy. *Annual Review of Ecology and Systematics* 33(1): 397–425.

Pearse, I.S., Koenig, W.D., Funk, K.A. and Pesendorfer, M.B., 2015. Pollen limitation and flower abortion in a wind-pollinated, masting tree. *Ecology* 96(2): 587–593.

Pigott, C.D., 1992. Are the distribution of species determined by failure to set seed. In *Fruit and Seed Production*. Cambridge University, Cambridge, pp. 203–215.

Pinheiro, F., Cafasso, D., Cozzolino, S. and Scopece, G., 2015. Transitions between self-compatibility and self-incompatibility and the evolution of reproductive isolation in the large and diverse tropical genus *Dendrobium* (Orchidaceae). *Annals of Botany* 116(3): 457–467.

Rasmussen, K.K. and Kollmann, J., 2004. Poor sexual reproduction on the distribution limit of the rare tree *Sorbus torminalis*. *Acta Oecologica* 25(3): 211–218.

Renner, S.S., 2014. The relative and absolute frequencies of angiosperm sexual systems: dioecy, monoecy, gynodioecy, and an updated online database. *American Journal of Botany* 101(10): 1588–1596.

Rieseberg, L.H., Hanson, M.A. and Philbrick, C.T., 1992. Androdioecy is derived from dioecy in Datiscaceae: evidence from restriction site mapping of PCR-amplified chloroplast DNA fragments. *Systematic Botany* 17(2): 324–336.

Rieseberg, L.H., Philbrick, C.T., Pack, P.E., Hanson, M.A. and Fritsch, P., 1993. Inbreeding depression in androdioecious populations of *Datisca glomerata* (Datiscaceae). *American Journal of Botany* 80(7): 757–762.

Saccheri, I., Kuussaari, M., Kankare, M., Vikman, P., Fortelius, W. and Hanski, I., 1998. Inbreeding and extinction in a butterfly metapopulation. *Nature* 392(6675): 491.

Sanzol, J. and Herrero, M., 2001. The 'effective pollination period' in fruit trees. *Scientia Horticulturae* 90(1–2): 1–17.

Sauquet, H., von Balthazar, M., Magallón, S., Doyle, J.A., Endress, P.K., Bailes, E.J., de Morais, E.B., Bull-Hereñu, K., Carrive, L., Chartier, M. and Chomicki, G., 2017. The ancestral flower of angiosperms and its early diversification. *Nature Communications* 8: 16047.

Shivanna, K.R. and Rangaswamy, N.S., 1992. *Pollen biology: a laboratory manual*. Springer Science & Business Media, Berlin.

Tanaka, N., Uehara, K. and Murata, J., 2004. Correlation between pollen morphology and pollination mechanisms in the Hydrocharitaceae. *Journal of Plant Research* 117(4): 265–276.

Tedder, A., Helling, M., Pannell, J.R., Shimizu-Inatsugi, R., Kawagoe, T., van Campen, J., Sese, J. and Shimizu, K.K., 2015. Female sterility associated with increased clonal propagation suggests a unique combination of androdioecy and asexual reproduction in populations of *Cardamine amara* (Brassicaceae). *Annals of Botany* 115(5): 763–776.

Wallander, E., 2001. Evolution of wind-pollination in *Fraxinus* (Oleaceae). PhD dissertation, Göteborg University, Sweden.

Wang, H., Yu, W.B., Chen, J.Q. and Blackmore, S., 2009. Pollen morphology in relation to floral types and pollination syndromes in *Pedicularis* (Orobanchaceae). *Plant Systematics and Evolution* 277(3–4): 153.

Watson, L. and Dallwitz, M.J., 2018. The genera of Leguminosae-Caesalpinioideae and Swartzieae: descriptions, identification, and information retrieval, version: 22, March 2017, available at https://www.delta-intkey.com/caes/fr/index.htm, accessed 21 December 2018.

Weeks, S.C., 2012. The role of androdioecy and gynodioecy in mediating evolutionary transitions between dioecy and hermaphroditism in the Animalia. *Evolution: International Journal of Organic Evolution* 66(12): 3670–3686.

Wilcock, C. and Neiland, R., 2002. Pollination failure in plants: why it happens and when it matters. *Trends in Plant Science* 7(6): 270–277.

Willmer, P., 2012. Ecology: pollinator–plant synchrony tested by climate change. *Current Biology* 22(4): R131–R132.

Wolf, D.E. and Takebayashi, N., 2004. Pollen limitation and the evolution of androdioecy from dioecy. *The American Naturalist* 163(1): 122–137.

Wolf, D.E., Satkoski, J.A., White, K. and Rieseberg, L.H., 2001. Sex determination in the androdioecious plant *Datisca glomerata* and its dioecious sister species *D. cannabina*. *Genetics* 159(3):1243–1257.

Wright, L.I., Tregenza, T. and Hosken, D.J., 2008. Inbreeding, inbreeding depression and extinction. *Conservation Genetics* 9(4): 833.

Yampolsky, C. and Yampolsky, H., 1922. Distribution of sex forms in the phanerogamic flora. *Bibliotheca Genetica* 3: 1–62.

Zar, J.H., 1974. *Biostatistical Analysis*. Prentice-Hall, Englewood Cliffs, NJ.

Zhang, Q., Onstein, R.E., Little, S.A. and Sauquet, H., 2018. Estimating divergence times and ancestral breeding systems in Ficus and Moraceae. *Annals of Botany* 123(1): 191–204.

Seed Biology and Modes of Regeneration

4

4.1 Introduction

Seeds are units of reproduction and contribute to the important roles of maintenance and continuity of plant communities on earth. These propagules play multiple roles in plant life cycles including being the beginning and bearers of the essentials of inheritance. Seeds represent units of multiplication and dispersal, continuation and innovation, survival, renewal and birth of angiospermic plants. These mysterious genetic capsules are the most important state for regeneration of sexually reproducing plants and have the potential to transform into an individual plant. Seeds contain the embryo – the new plant in miniature – and are structurally and physiologically equipped to become a dispersal unit. Seeds are also well provided with food reserves to sustain the growing seedlings until they establish as a self-sufficient, autotrophic organism. In nature, seeds pass from a plant through a very temporary environment to a place on or in the ground where they will probably lie dormant for a while and then germinate at the onset of favourable conditions to grow and establish into new plants. Seeds may also disappear due to predation, extreme weather conditions, damage by pathogens, natural death due to ageing and germination with or without emergence.

As seeds are the most critical stage in the life cycle of a plant, they play a vital role in the continuity of a plant species. However, the behaviour of seeds varies widely across species. It is therefore essential to study the ecological requirements of seeds for conservation and rehabilitation purposes as well as silvicultural and forest management plans (Guariguata, 2000; Nambara and Nonogaki, 2012). Species differ in their reproductive traits such as the minimum size at which they start to reproduce (van Ulft, 2004), frequency of reproduction (Janzen, 1978), quantity and size of seed produced (van Ulft, 2004), dispersal mechanism and dispersal distance (Willson, 1993), as well as among-population variation (Cochrane *et al.*, 2015). Various environmental factors differ among habitats, such as temperature, humidity, light, soil characteristics, dispersal syndromes, germination time and densities of competing plants. Herbivores and fungi affect production and selection for different seed sizes (Khurana and Singh, 2001). Seed predation by insects, birds, mammals and damage by pathogens are important factors that profoundly affect plant population dynamics and regeneration of plant species (Maron and Crone, 2006; Bagchi *et al.*, 2014). Pathogens are an important cause of mortality for tropical seedlings and their affect is aggravated in moist, heavily-shaded habitats (Augspurger and Kelly, 1984), as well as regulating the rainforest plant diversity and composition (Bagchi *et al.*, 2014). On the other hand, attraction of fruits and seeds by mammals, birds and insects helps in dispersal of different kinds of fruits at different degrees (Roosmalen, 1985; Brewer and Rejmanek 1999; Mello *et al.*, 2015) and involves reciprocal benefits to the animals with nutritional rewards (Vander Wall and Beck, 2012; Mello *et al.*, 2015). According to some theories, the dispersal of fruits or seeds away from the parent tree can enhance their chance of arriving at and surviving in a suitable micro-environment and escaping mortality near the parent (Janzen, 1970; Howe and Smallwood, 1982; Hughes *et al.*, 1994; Jansen *et al.*, 2014).

Viability and germination are two important events in the life cycle of a plant. On the basis of viability, seeds are classified into two categories: (a) orthodox seeds which are usually dormant and can retain their viability for longer periods even under fully hydrated oxygenic conditions; and (b) recalcitrant seeds, which have little or no dormancy and cannot be dried below a critical moisture content (Roberts, 1973). In general, orthodox seeds are generally dried to low moisture contents and stored in 'conventional' storage in a freezer at −20°C, while recalcitrant seeds must be preserved cryogenically in liquid nitrogen (LN) for storage (Walters *et al.*, 2013).

Seeds may have evolved thick seed coats as defensive structures for preventing penetration of insect ovipositors. The phenolic compounds in the seed coat contribute to the inhibition of microorganisms (Janzen, 1969; Mohamed-Yasseen *et al.*, 1994). Recent studies have shown that the seed coat also acts as a channel for transmitting environmental cues to the interior of the seed to adjust its metabolism in response to changes in its external environment (Radchuk and Borisjuk, 2014). On the other hand, seed dormancy, defined as the incapacity of an intact viable seed to complete germination under favourable conditions (Finch-Savage and Leubner-Metzger, 2006; Footitt *et al.*, 2011) is important to prevent germination of seeds during unfavourable periods for seedling growth and establishment. Different degrees of dormancy in plants also assist their next generation to grow and develop across time and ensure survival in unpredictable environmental conditions (Poisot *et al.*, 2011). Several strategies have been developed by plants to regulate seed germination to maximize survival and establishment of seedlings in local climatic conditions. Seed dormancy is a very important feature of plants and has played a critical role in their survival and evolution over millions of years.

Seed dormancy has been primarily classified into five different categories: physiological (PD), morphological (MD), morphophysiological (MPD), physical (PY) and combinational (PY + PD) (Baskin and Baskin, 1998, 2004). PD is a programmed state of seeds that restricts the set of environmental conditions such as temperature, water potential, light spectra, light intensity, photoperiod and related chemical cues (Nelson *et al.*, 2012). MD is usually displayed in seeds in the form of underdeveloped (in terms of size) but well differentiated embryos which eventually grow and germinate with time (Jacobsen and Pressman, 1979). MPD displays extra physiological components to MD traits (Baskin and Baskin, 2004). PY dormancy is primarily due to a physical barrier to water or similar liquid substances and is displayed in the form of thick layers of palisade cells in the seed or fruit coat (Baskin, 2003; Baskin and Baskin, 1998). Combinational dormancy is a combination of PY and PD types and shows traits such as water-impermeable coats and physiological embryo dormancy (Baskin and Baskin, 2004).

Both morphological and physiological conditions of the seed and the embryo are important determinants of seed germination (Bewley *et al.*, 2012). Recent studies show epigenetic regulation of seed dormancy and germination (reviewed in Nonogaki, 2017). Seed-coat induced dormancy may pose difficulty when developing regeneration strategies and can be broken artificially by mechanical scarification, hot water, dry heat or acid or other chemical treatments. Successful germination involves complex interactions between the seed's physical position, its physiological state and its micro-environment (Hamrick and Lee, 1987).

Information on the ecological requirements at the seed and seedling stages is vital both for conservation and rehabilitation but is seldom articulated in silvicultural and forest management plans (Guariguata, 2000; Khurana and Singh, 2001). Knowledge of the ecology of germination and seedling growth is critical, not only for understanding the community processes of plant recruitment and succession, but also for developing strategies for conservation of biodiversity and restoration of tropical forests (Khurana and Singh, 2001). Seed germination of endangered species at the right time and space is not only important for the maintenance of regeneration but also for the species' ex-situ conservation (Herranz *et al.*, 2010).

4.2 Materials and Methods

4.2.1 Pod collection and measurement of pod traits

In this study, pods were collected from seven mature *G. assamicus* trees representing the three sites (Dambla village, Moishing and Yewang village) during January–February when pods were mature and ready for shading. Pods were transported to the laboratory for measurement of different quantitative traits and conduct other seed biology experiments. A composite collection of 100 pods from three sites was measured for length, width and breadth using digital calipers (Mitutoyo, Japan). The same set of pods were then opened to count the number of seeds per pod to construct a fruit size distribution curve.

This was compared with log-normal distribution and tested for probability with Kolmogorov-Smirnov test (STATISTICA version 6.0).

4.2.2 Seed dimensions and weight

Pooled samples representing all seven trees were used for measuring seed dimension and seed-fresh weight. Dimensions of seeds were measured with the help of digital calipers (Mitutoyo, Japan). A total of 257 freshly collected seeds were weighed individually in an electronic balance to generate a seed mass distribution curve and were compared with a normal distribution following the Kolmogorov-Smirnov test (STATISTICA version 6.0).

4.2.3 Seed morphology and anatomy

Seed morphology (i.e. shape, colour, hardiness and surface features) was examined visually. Seed surface features and internal structures like dermal layers, cotyledons and embryo were observed under a stereo microscope. Scanning electron microscopic (SEM) observations were made to study the fine details of seed surface features and were interpreted based on available literature. For SEM, small sections of dried seed coats were excised, mounted on metal stubs following gold coating and observed under an SEM (LEO 1430VP, Karlzeiz, German). Digital photographs were taken under different magnifications for further examination.

4.2.4 Seed viability, germination and moisture content

Seed dormancy and loss of viability were tested at three-month intervals for a year. Seeds were stored in paper bags under normal laboratory conditions (inside wooden cabinets in complete darkness; temperature 20 ±5°C) until they were used. Seed viability was assessed using the tetrazolium test (MacKay, 1972). In this test, red staining due to formation of red-coloured formazan in the embryo tissue is considered as an indicator of metabolic activity in the resting seed. Staining was done with 1% (W/V) freshly prepared aqueous solution of 2, 3, 5-triphenyl-tetrazolium chloride (pH 7.0) at room temperature (25–27°C). Seeds were first imbibed in tap water; cotyledons were then excised and placed in the test solution for 24 h in the dark and scored as stained or unstained. Germinability was tested by placing seeds in Petri dishes over moist filter paper kept in normal daylight conditions at room temperature. Germination was defined as the emergence of radicle from the seed coat. Newly-germinated seeds were counted and removed from the dish. Observations were made for 60 days, after which the overall germination rate was calculated. Tests were carried out in laboratories at the campus of North Eastern Regional Institute of Science and Technology (27°07′ N latitude, 93°22′ E longitude, 110 m asl). Moisture content of the seeds was also estimated at three-month intervals to asses the rate of loss of seed moisture during storage. For this, a fresh weight of 50 seeds was taken using a digital balance following oven-drying at 100°C for 48 h. Oven-dried seeds were re-weighed and the percentage moisture content calculated.

4.2.5 Seed dormancy and different treatments to break dormancy

Seeds of *G. assamicus* are extremely hard and it was predicted that seed coat-imposed dormancy might be primarily responsible for delayed germination. In this study, both mechanical and chemical treatments were given to the seeds to break seed-coat imposed dormancy and develop suitable strategies for silvicultural practices. The treatments were (1) control (untreated seed); (2) mechanical scarification (the seed coat was sanded with wood sandpaper opposite the embryo until the cotyledon was exposed); (3) hot water treatment (soaking the seeds in boiling water for 30 seconds, 1, 2, 3 and 5 minutes; and (4) chemical scarification of intact seeds with concentrated sulphuric acid for 2, 5, 10, 20, 40, 60 and 120 minutes followed by washing in tap water for 10 minutes. Seeds from each treatment were germinated in Petri dishes with moist filter paper under normal laboratory conditions (temperature 20 ±5°C and about 12-hour photoperiod).

4.2.6 Seed germination at different soil depths

Seed germination was tested in polythene bags filled with farmyard soil (pH 5.4, nitrogen 0.3%, organic matter 4.3%) keeping them in a mist chamber under an agro-shed net which protected them from direct sunlight. Seeds were sown at different soil depths (0 cm, i.e. surface, 1 cm and 2 cm) to infer the effect of soil depth on seed germination. Watering was done regularly to keep the soil moist throughout the experimental period. New germination was recorded on an everyday basis and continued for 60 days, after which cumulative germination percentage was calculated.

4.2.7 Seed dispersal and predation in the field

Seed dispersal, predation and damage by various agents were observed during peak fruiting season in the field and by interviewing local people. As the *G. assamicus* tree is intricately associated with the culture and tradition of the region, local people show enormous interest in many facets of the life history of the species, including seed dispersal and predation. Dispersal by natural means was also observed during the study period. Pathogen-infected seeds from the forest floor were brought to the laboratory and pure cultures were developed. Fungal pathogens were identified by staining the mycelia from pure cultured colonies using suitable stains and were observed under a microscope. Identification was done by following manuals and available literature.

4.2.8 Vegetative mode of regeneration

Terminal parts of the branches of same diameter were cut into 150 mm long pieces using sharp pruning shears. Leaves from the lower half of the cuttings were completely removed and partially trimmed from the upper half to minimize water loss. In this experiment, 100, 200 and 500 ppm (parts per million) of indole-3-acetic acid (IAA), indole-3-butyric acid (IBA), 1 naphthaleneacetic acid (NAA) and 2,4-dichlorophenoxyacetic acid (2,4-D) were used. The basal ports of the cuttings (about 1.5 cm) were soaked carefully in hormone solution for one minute by maintaining the vertical orientation. The treated cuttings were then planted into polythene bags filled with farmyard soil and transferred to a partially shaded nursery for rooting. Regular watering was done to maintain constant soil moisture levels for three months. This experiment was conducted during September/October at Dirang and was monitored regularly for signs of rooting, leafing and/or branching.

Statistical analysis was done following Zar (1974) and using software packages (MS Excel, SYSTAT, STATISTICA and ORIGIN).

4.3 Results and Discussion

4.3.1 Pod and seed characteristics

Mature *G. assamicus* pods are 6–17 cm long having 23–35 x 18–23 mm dimensions. Number of seeds per pod ranges from 1–8 and most of the pods (60%) are 5–7 seeded (Fig. 4.1). Frequency distribution of seeds per pod is almost normal (Fig. 4.2). Generally, a species may exhibit unique patterns of frequency distribution of seed number per fruit and range from highly positive skewed, to normal, to negatively skewed distribution (Shaanker *et al.*, 1988; Ganeshaiah and Shaanker, 1992). While species with positively skewed distribution show dispersal mechanism (usually through wind, water or animals) (Ganeshaiah and Shaanker, 1992), seeds of the species which show negatively skewed distribution represent the units of dispersal themselves as well as demonstrating low rates of seed abortion (Ganeshaiah *et al.*, 1986; Shaanker *et al.*, 1988). Factors such as decrease in pollen quality and reduced pollination intensity in small populations largely contribute to the low fruit and seed production in smaller populations (Lee, 1988; Kwak *et al.*, 1998; Jacquemyn *et al.*, 2001). In the case of *G. assamicus*, almost normal distribution of seeds per pod indicates very little or no such limitations in its natural habitat. Experiments on breeding systems and extensive pollination experiments did not show any strong evidence of pollen limitation in *G. assamicus* (Choudhury *et al.*, 2014).

Number of seeds produced per mature tree represents one of the most important basic parameters of the population dynamics of a species (Kelly, 1984). In general, an individual plant would achieve its maximum fitness only if it produced an infinitely large number of big seeds on a regular basis (Primack, 1987). While seed production largely depends on the availability of resources for reproduction and the amount used in the production of a certain number of seeds (Primack,

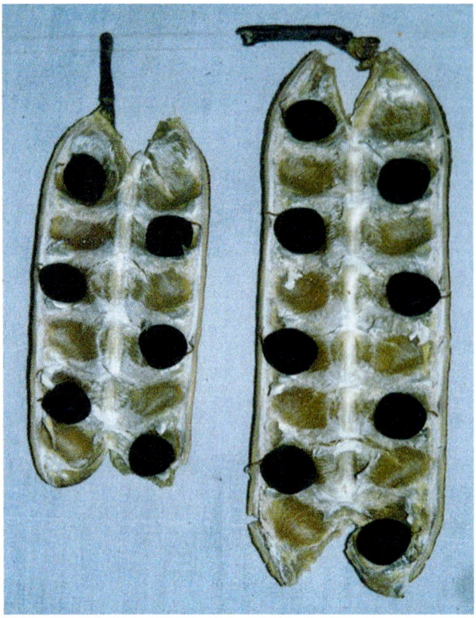

Fig. 4.1. Mature *G. assamicus* pods showing attached seeds (photo: Baharul I. Choudhury).

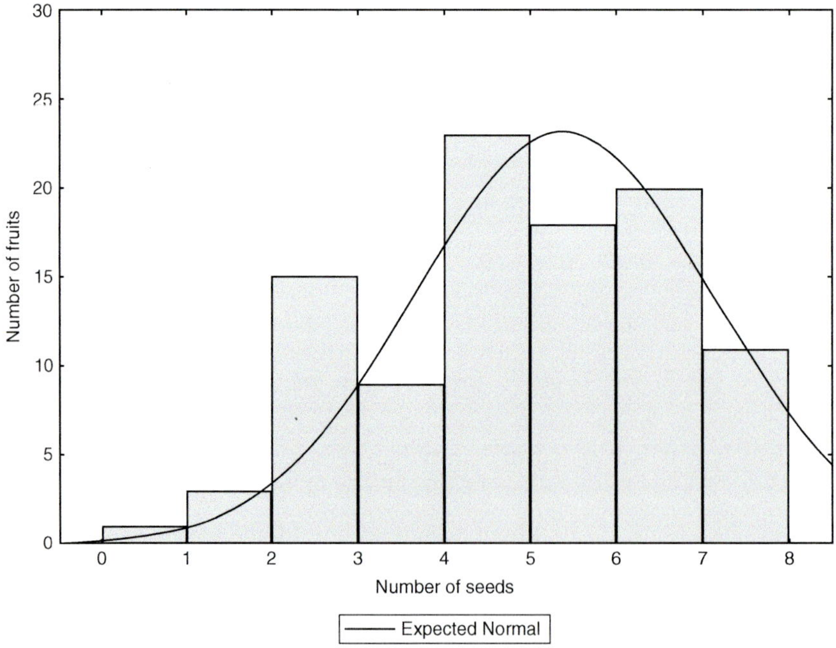

Fig. 4.2. Frequency distribution of *G. assamicus* seeds per pod.

1987), seed size may be achieved through the compensation of producing fewer seeds per fruit and fewer fruits per plant (Primack, 1978). Overall, the dimension of *G. assamicus* pods and number of seeds per pod collected from the three sites did not show any variation. This may be due to similar physiography and age structure of the individual trees in all the sites (Primack, 1987; Paul-Victor and Turnbull, 2009).

Seeds are 17.24 mm × 15.21 mm in dimension and are extremely hard. Seed weight varied from 1.2–2.4 gm per seed and the majority of the seeds had a weight range between 1.6–2.2 gm. Seed size and mass directly influence seed dispersal and seed predation at the seed stage, as well as seedling establishment, growth, and survival at the later stages and play crucial ecological roles in their long-term continuity (Fenner, 1983; Michaels *et al.*, 1988; Larson *et al.*, 2015). Large seed size in general demonstrates optimal performance in terms of germination rates, seedling growth and shoot length over the medium and small seed class (Khan and Shankar, 2001; Mishra *et al.*, 2014). It is established that single-seed mass in today's plant species may vary over a range of 11.5 orders of magnitude (Moles *et al.*, 2005). In the case of *G. assamicus* seed weight, the range is comparatively much smaller. Seed weight distribution was found to be slightly asymmetrical and negatively skewed in the Kolmogorov-Smirnov (KS) test ($d=0.093078$, $p=0.01$) having majority of seeds in the heavier weight class (Fig. 4.3). Seed mass is under strong genetic control (Paul-Victor and Turnbull, 2009) and a plant's ability to modify the seed size from smaller to larger may be associated with the adaptation to newer ecological conditions (Igea *et al.*, 2017).

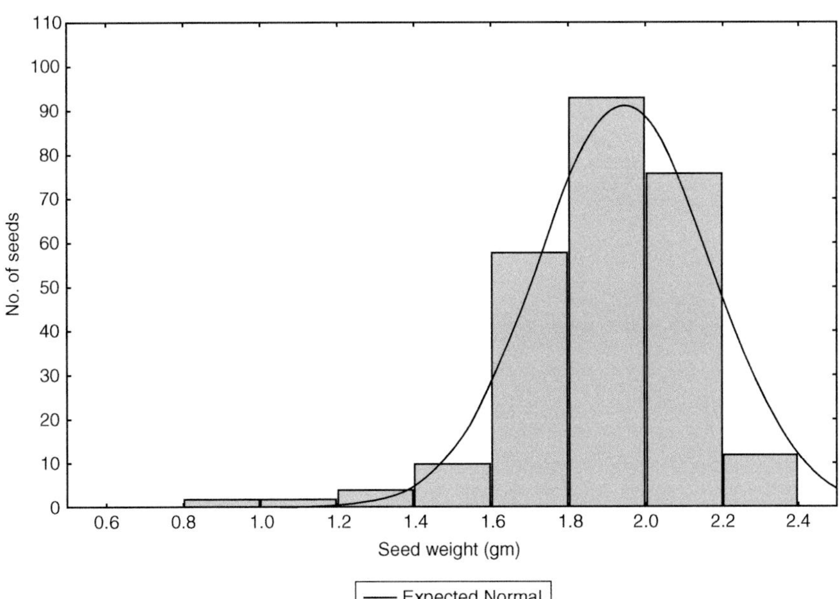

Fig. 4.3. *G. assamicus* seed weight distribution based on KS test.

4.3.2 Seed morphometry and related properties

G. assamicus seeds are ovoid in shape and the surface is fine-textured (Fig. 4.4). The outer black testa is extremely hard and impermeable in nature. The testa in seeds is a layer of the embryo that protects against severe environmental conditions. Hard testa primarily control germination by imposing dormancy and by limiting other physical and biological agents during unwanted ecological conditions (Weber *et al.*, 1996; Debeaujon *et al.*, 2000). More importantly, seed coat-imposed dormancy is part of the seed survival strategy of several plant species (Kelly *et al.*, 1992). A few studies have suggested correlation between seed coat colour and dormancy due to phenolic compounds in different species. For example, dark seeds of proso millet (*Panicum miliaceum* L.) showed thicker seed coats and consequently imbibed and germinated more slowly, thereby suffering less imbibition damage, and persisting longer in soil than light-coloured seeds (Khan *et al.*, 1997).

Seeds of *G. assamicus* are dark black and might be rich in similar chemical properties to resist water imbibition and delay germination. However, cracks on the seed surface are easily visible under a dissecting microscope (Fig. 4.5). There

Fig. 4.4. Mature *G. assamicus* seeds (photo: Baharul I. Choudhury).

Fig. 4.5. Visible cracks on seed coat under dissecting microscope (photo: Baharul I. Choudhury).

is a stony endodermal layer (when dry) inside the testa (Fig. 4.6) which becomes mucilaginous after imbibition (Fig. 4.7). Seed surface morphology reveals that a hard and impermeable testa is the main barrier for imbibition and consequently retards germination. Hard-coat induced seed dormancy and subsequent delayed germination varies considerably in different species and is mostly attributed to limited water uptake and gas exchange, presence of chemical inhibitors, barriers against the escape of inhibitors from the embryo, modification of the light penetration to the embryo and mechanical restraint of the embryo (Bewley and

Fig. 4.6. Stony-hard testa (indicated by arrows) of mature seeds (photo: Baharul I. Choudhury).

Fig. 4.7. Hard testa turns mucilaginous after imbibition (photo: Baharul I. Choudhury).

Black, 2012). Moreover, hard-coat dormancy is a common feature in legumes (Quinlivan, 1971; Baskin and Baskin, 1989; Thapliyal *et al.*, 1998; Smýkal *et al.*, 2014). Unlike most other legume species, *G. assamicus* is distributed in the temperate regions (Menon *et al.*, 2010) and might have developed coat-imposed dormancy to prevent extreme ecological conditions and maintain the soil seed bank for as long as possible. Such dormancy in legumes often imposes low germination rates (Abbo *et al.*, 2009) and is a challenge for management and restoration of such species in a natural habitat.

Extremely hard seed coat in the mature dry state is generally due to the presence of heavily thickened galactomannan or mannan polymers on the walls of the endosperm cells, well observed in carob (*Ceratonia siliqua* L.) (McCleary and Matheson, 1974) and fenugreek (*Trigonella foenum-graecum* L.) (Reid, 1971). Galactomannans are found in the endosperm of several legume seeds and plays the role of reserving polysaccharide and maintaining a moist environment for the germinating seed (Dea and Morrison, 1975). Galactomannan is also present in developing seeds of *Gleditsia triacanthos* (McCleary *et al.*, 1987). *Gleditsia triacanthos* is a sister genus of *G. assamica* and therefore strongly indicates its presence in the seeds of the latter. Some have suggested that the emergence of embryos from hard seeds is most likely associated with the cell wall structure of the micropylar endosperm and not to endo-β-mannanase activity (Gong *et al.*, 2005). The endosperm walls contain hydrophilic galactomannans which become mucilaginous after a period of imbibition and hydrolysis (Reid and Bewley, 1979). It was observed that seeds of *G. assamicus* transform into mucilaginous coverings after imbibition and it is possible that galactomannan or mannan polymers are present in the endosperm wall. Hard seed coats with a palisade or malpighian layer in the Leguminosae and Malvaceae families are due to special devices which enable seeds to imbibe water preceding germination. Seed coats only become permeable after release or splitting of a special region of the malpighian layer (Graven *et al.*, 1997).

G. assamicus seeds have two cotyledons with a basal embryo. Relatively tiny embryos lie within the cotyledons (Fig. 4.8) in a hollow cavity of the seed (Fig. 4.6). SEM observation showed highly compact outer testa with marked cracks (Fig. 4.9). Cracks were more in the vicinity of the micropyle than the other areas. On the other hand, SEM at higher magnification revealed a densely cuticulated surface throughout the seed (Fig. 4.10). Compact and thick testa may be primarily preventing uniform intake of water throughout the seed surface. This may have led the *G. assamicus* seeds to be highly impermeable in nature. More cracks in the vicinity of the micropyle promote imbibition as well as rapid activation of the embryo. SEM studies also revealed the seed coat of *G. assamicus* as having a continuous layer of tightly packed palisade cells. Hard-seededness due to tightly packed palisade cells in the seed coat might be the major barrier to water entry into seeds (Cavanagh, 1987; Egley, 1989; Baskin and Baskin, 2014). Though seed coat impermeability varies in most leguminous species, the lens (strophiole) forms a structural weakness which can be disrupted naturally or artificially, thus rendering the seeds permeable to water (Teketay, 1997).

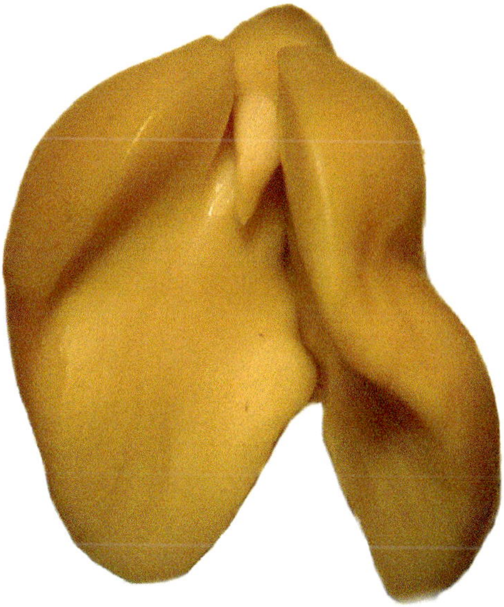

Fig. 4.8. *G. assamicus* embryo protected by the cotyledons (photo: Baharul I. Choudhury).

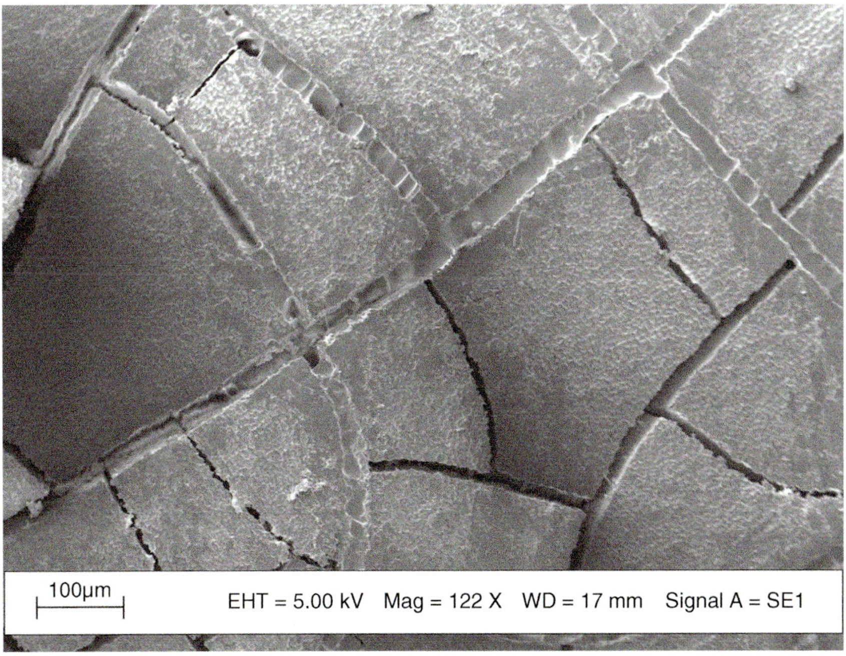

Fig. 4.9. SEM of seed surface showing prominent cracks (photo: Baharul I. Choudhury).

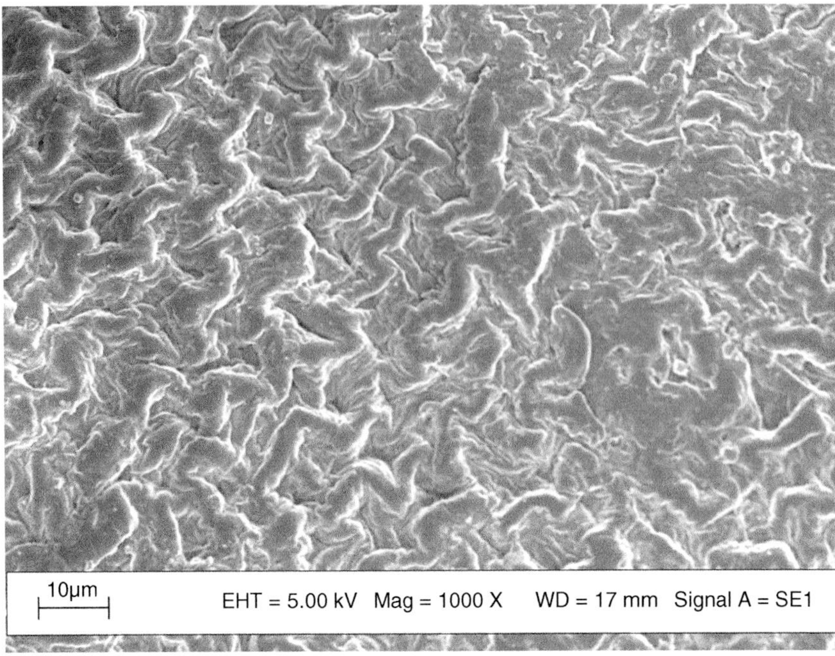

Fig. 4.10. Densely cuticulated surface throughout the seed coat (photo: Baharul I. Choudhury).

4.3.3 Seed viability, germination and moisture content

Seeds showed a very high percentage (90%) of viability immediately after harvesting. A high percentage (46%) of seeds was viable even after 12 months of storage at room temperature in normal laboratory conditions (Fig. 4.11). Germination percentage of fresh seed was also high (42%) which gradually decreased to 6.7% after a year of storage (Fig. 4.11). *G. assamicus* seeds were viable for a long period of time and such seed types are, as previously noted, referred to as orthodox (Choudhury *et al.*, 2009). Loss of seed moisture content at three-month intervals showed a meagre change over a one-year study period (Table 4.1) indicating copious viability for longer duration. Such seeds with high initial viability and germination percentage retain their viability for longer than immature seeds (Stein *et al.*, 1974). These observations suggest very easy and long-term preservation of mature seeds for successful seed germination and seedling raising in the future.

4.3.4 Different treatments to break dormancy

Artificially breaking dormancy of orthodox seeds requires appropriate treatments to enhance seed germination, and nursery development. Several studies showed significant improvement in terms of seed germination, and 'dormancy breaking'

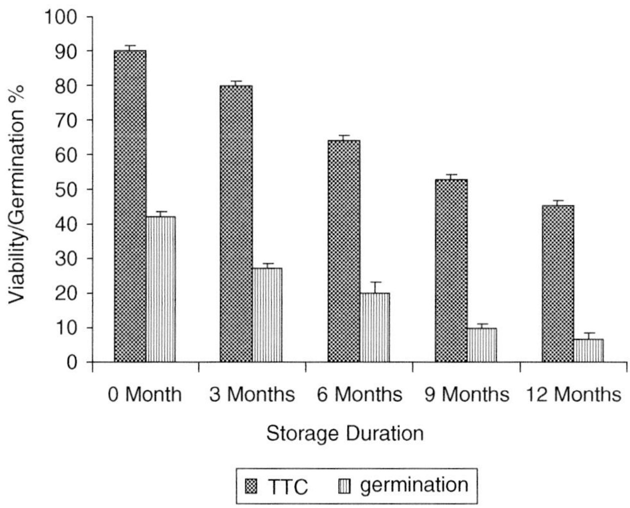

Fig. 4.11. Seed viability of *G. assamicus* at normal laboratory storage condition (bars represent ± standard error).

Table 4.1. Status of *G. assamicus* seed moisture content during one-year storage period at room temperature in normal laboratory conditions (± standard error).

Months of storage	0 months	3 months	6 months	9 months	12 months
Moisture content	21.95 (±0.94)	17.39 (±0.88)	15.90 (±1.37)	14.91 (±0.76)	11.35 (±1.25)

is a popular practice, especially in legumes (Herranz *et al.*, 1998; Sy *et al.*, 2001; Baskin and Baskin, 2004; Bhatt *et al.*, 2016). Pre-treatment of *G. assamicus* seeds by mechanical scarification, hot water treatment and chemical scarification with sulphuric acid improved germination at different degrees. Boiling water treatment for two minutes dramatically increased seed germination percentage by almost double (80%) compared to the control (42%). Other treatments such as mechanical and chemical scarification also showed significant improvement in rapid and uniform germination as compared to the control (Fig. 4.12).

Very hard testa was found to be the main barrier for impermeability in the case of *G. assamicus*. Seed coat impermeability is related to hard seed coat as well as presence of a waxy layer surrounding the cotyledons. The degree of seed coat thickening is influenced by genotype, conditions during the growing season and dehydration rate of the seed (Patane and Bradford 1993; Russi 1993). Results from the present study showed that seed coat-imposed dormancy due to water impermeability in *G. assamicus* can be broken by mechanical scarification, boiling water or acid treatments (Fig. 4.12). This result indicates that dormancy in *G. assamicus* seeds is primarily of PY type due to water-impermeable layers of palisade cells

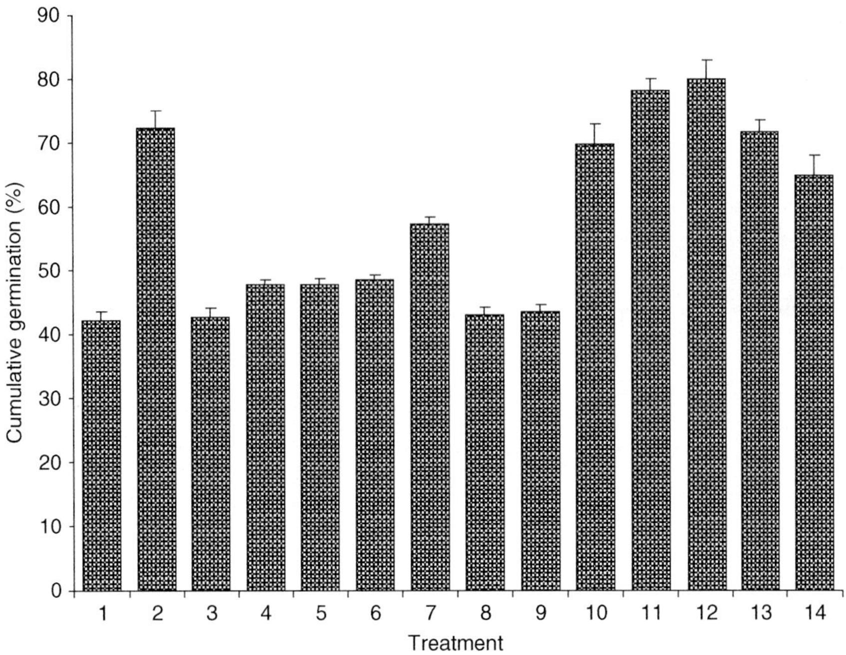

Fig. 4.12. Effect of various seed treatments (mechanical scarification, sulphuric acid (SA) and boiling water (BW)) of *G. assamicus* seeds to improve seed germination. 1 = control; 2 = mechanical scarification; 3 = SA for 2 minutes; 4 = SA for 5 minutes; 5 = SA for 10 minutes; 6 = SA for 20 minutes; 7 = SA for 40 minutes; 8 = SA for 60 minutes; 9 = SA for 2 h; 10 = BW for 30 seconds; 11 = BW for 1 minute; 12 = BW for 2 minutes; 13 = BW for 3 minutes; and 14 = BW for 5 minutes (bars represent ± standard error).

of the seed coats (Baskin, 2003). Similar results have also been demonstrated by several researchers in different legume species (Larsen, 1964; Khan and Tripathi, 1987; Danthu *et al.*, 1992; Masamba, 1994). Interestingly, boiling water treatment improved germination more effectively than other treatments. Boiling water treatment also showed significant improvement in seed germination of another legume, *Acacia negrii* (Teketay, 1997). However, some legume species responded poorly to boiling water treatments (Larsen, 1964; Clements *et al.*, 1977; Danthu *et al.*, 1992). Such relatively simple, convenient and economic methods which require little skill can also be practically applied to large-scale seedling production and restoration measures.

Seed coat impermeability, which is a very common characteristic in most legumes, is a delaying mechanism and prevents germination under unfavourable conditions (Barton, 1965; Tran and Cavanagh, 1984; Baskin and Baskin, 1989; Tybrik, 1991; Bouman and Boesewinkel, 2017). Such hard seed coat also helps seeds to survive in the soil because seeds must not only remain viable, but also not germinate before a favourable period. Hard seed-coat imposed dormancy is

Fig. 4.13. *G. assamicus* pods partially eaten by grazing animals in the field (photo: Baharul I Choudhury).

common in perennial leguminous plants and may share up to 80% and more (Travlos *et al.*, 2007; Smýkal *et al.*, 2014). Orthodox seeds acquire desiccation tolerance during development and may be stored in the dry state for predictable periods under defined conditions. Hard seed coat also allows endozoic dispersal, recolonization after fire and helps the seeds to withstand unfavourable conditions such as heat, drought, digestive juice as well as mechanical damage (Tybrik, 1991). *G. assamicus* seeds were also found ingested by grazing animals (Fig. 4.13) and there are possibilities that passage through a digestive tract improves germination and thus facilitates better regeneration in the natural habitat.

Pods are also a highly preferred diet of wild animals such as deer and wild boar. Local tribes often used fruiting *G. assamicus* trees to hunt wild animals when they visited the vicinity to eat mature pods during fruiting season (personal observation). This was a popular hunting strategy in the area when wild animals were abundant. However, such strategies are unproductive now due to the rapid decline of wild animals in addition to forest degradation and expansion of human habitat in forest areas.

Seed pre-treatment in the laboratory reflects the ways by which *G. assamicus* seeds overcome dormancy in nature. Permanent action of weak acidic soil in the study site softens the seed coat, as presumed from laboratory treatment in strong acid. Chewing and ruminating the pods by animals may provide scarification of seeds and improve imbibition as well as germination. Various pre-treatment methods of *G. assamicus* seeds showed significant improvement in rapid and uniform germination. However, boiling water treatments for two minutes may be highly recommended for its low cost and ease of use.

4.3.5 Seed germination at different soil depths

Several microsite characteristics such as light intensity, temperature and burial depth affect seed germination and emergence at different degrees (Dalling *et al.*, 1995; Benvenuti *et al.*, 2001; Jusaitis *et al.*, 2004). Seed burial depth may act as a critical requirement for seed germination and successful establishment of seedlings (Limón and Peco, 2016; Kolář *et al.*, 2017). Seed germination in *G. assamicus* was hypogeal as the cotyledons remain below ground after germination. Germination at different soil depths showed different results. Maximum percentage of seed germination (42%) was observed when seeds were sown at 1 cm soil depth while it was minimum (20%) at 2 cm depth. There was significant variation in seed germination percentage among different soil depths ($F=13.038$; $p<0.05$). Germination experiments at different soil depths indicated that *G. assamicus* seeds germinate well at 1 cm depth rather than surface or 2 cm depth. Similar results were also obtained from other legume species (Bradstock and Auld, 1995; Williams *et al.*, 2004). It has been suggested that larger seeded species could benefit the most from increased temperature penetration into the topsoil, because their larger starch reserves might allow survival of germinants from greater depths in the topsoil (Midgley and Bond, 2001). *G. assamicus*, being a legume, has seeds with rich nutrient content and might have developed a strategy for survival at greater depths.

Seeds remain viable for a year and the estimation revealed that 80% of seeds remain viable after three months (Fig. 4.11), i.e. during June–July when outdoor conditions are most favourable for seed germination. While seed germination with respect to soil depth showed that *G. assamicus* germinates best at 1 cm depth, better germination for the New South Wales legume *Acacia suaveolens* was also recorded from medium soil depth (Bradstock and Auld, 1995). This may be an advantage to escape predation and damage by pathogens as well as washout by rainwater.

4.3.6 Dispersal

Seed dispersal is the physical movement of mature seeds from their mother plant to a different location to begin a new life cycle. Seed dispersal is one of the most important components of plant life cycles that determines the spatial arrangement and physical environment of seeds (Howe and Smallwood, 1982; Herrera *et al.*, 1994). As the seeds are of orthodox type, the species may be designated as 'fugitive' in nature (Hutchinson, 1951) and good dispersal mechanisms would play an important role in permitting the species to colonize new habitats (Salisbury, 1942).

4.3.7 Seed loss, predation and damage

Seed predation by various animals is common in the majority of forest tree species. This mechanism plays a crucial role in the life cycle of a species and consequently the forest community in general. Seed predation largely determines the maintenance of a forest ecosystem and has been a major selective force in regulating different aspects of seed biology, flowering, fruiting and dispersal behaviour (Janzen, 1969; Harper *et al.*, 1970; Brown *et al.*, 1979). Mature *G. assamicus* pods are fleshy and very rich in saponin content (Choudhury *et al.*, 2007). Stony, hard seeds are also very rich nutritionally and attract wild animals such as deer and wild boar as well as domestic animals like hill cattle and goats. Such activities greatly influence seed dispersal and have important implications for the population dynamics of the plant species involved (Gardener *et al.*, 1993). Ingested seeds may also escape mastication processes by animals and may be regurgitated or survive the passage through the digestive tract and be deposited with faeces that facilitate germination (Archer and Pyke, 1991 [Archer and Smeins 1991 in refs]). Similar activities by herbivores indicate the potential role in dispersal of *G. assamicus* seeds in the natural population. After ingestion, seeds are subjected to strong acidic action in the digestive tract, high temperatures and rumen microbes which dramatically increase germination rates, especially for hard seeds (Gardener *et al.*, 1993). Survival of legume seeds after ingestion by cattle has been associated with coat hardness (Gardener *et al.*, 1993; Miller and Coe, 1993; Venier *et al.*, 2012). On the other hand, the possibility of seed damage due to mechanical and chemical actions in the process of endozoochory is also high. Studies suggest that the probability of seed survival after ingestion by herbivores is related to seed mass, shape and coat thickness (Peco *et al.*, 2006; Traveset *et al.*, 2007). *G. assamicus* seed size, shape and hardness suggest higher possibilities of survival after passing through herbivore digestive systems.

During the present study, pools of intact and partially eaten *G. assamicus* pods and seeds were found on tree cavities (Fig. 4.14). Frugivory by arboreal animals is common in many forest species and influences the seed dispersal mechanisms both positively and negatively (Janzen *et al.*, 1985; Chanthorn *et al.*, 2018). Partially damaged *G. assamicus* seed clearly indicates seed-hoarding activities and at least some degree of seed damage by arboreal animals, most likely by rodents or squirrels. Although animals play a significant role in dispersal, especially by the frugivorous vertebrates through endozoochory, seed damage is a common phenomenon and has been observed in many species including legumes (Campos and Ojeda, 1997; Benítez-Malvido *et al.*, 2003).

Fully mature *G. assamicus* pods fall from the tree by gravity and come in contact with soil. Dry pods subsequently absorb moisture from the soil, snowfall or from precipitation providing an excellent media for microbial growth. As a result, pods on the forest floor were found largely infested by fungi, mainly *Aspergillus* sp., *Mucor* sp. and *Humicola* sp (Fig. 4.15). Dung beetles were also found foraging on *G. assamicus* pods on soil seed banks (Fig. 4.16). Long-term fungal infection leads

Fig. 4.14. *G. assamicus* pod partially ingested by grazing animals (photo: Baharul I. Choudhury).

Fig. 4.15. *G. assamicus* pods infested by fungi in field conditions (photo: Baharul I. Choudhury).

Fig. 4.16. Seed damage by insects (highlighted by circle) in the field (photo: Baharul I. Choudhury).

to seed damage which contributes significantly to the damage of mature fruits in the soil. Our seed germination trials in the greenhouse also demonstrated severe damage by fungal pathogens, reflecting natural process in field conditions. Seed damage by pathogens also influences the species composition and forest community structure (Spear *et al.*, 2014). Fruit rot due to *Alternaria alternata* has been reported in citrus, apple and pear (Domsch *et al.*, 1993) and in *Pinus densiflora* and *Cryptomeria japonica* seeds (Watanabe, 2002). Seed rot by forest pathogens significantly inhibits natural regeneration of several species at the seed stage (Cho *et al.*, 2007). Fungi play an important role in seed survivorship in soil (Crist and Friese 1993; Leishman *et al.* 2000; Blaney and Kotanen, 2001), and many fungi inhibit seed germination (Dinoor and Eshed, 1984) causing severe damage to the soil seed banks. In the present study, infection rate by *Aspergillus* and *Mucor* sp. was found highest in *G. assamicus* pods retrieved from the field. As the size of a single species population is related to seed supply and largely determined by favourable conditions available for germination and establishment (Harper, 1979), the major constraints for extremely small populations of *G. assamicus* may be due to poor seed supply, dispersal and poor germination.

4.3.8 Vegetative mode of regeneration

Vegetative propagation is an alternative method of rapid regeneration of plant species, especially for species of economic and ecological importance. This approach has been tried for different woody and non-woody plant species and found to be successful in conserving species of interest (Ofori *et al.*, 1996; Ratnamhin *et al.*, 2011; Wadl *et al.*, 2014). Our trial using different root stimulating hormone treatments for the vegetative mode of propagation did not show signs of root initiation during the experimental period. Though the branch cuttings showed leaf initiation at the beginning, they eventually died spontaneously. A protocol for in-vitro regeneration of *G. assamicus* has also been developed which showed promising results (Gupta *et al.*, 2013) and may be implemented for rapid and mass regeneration of the species.

4.4 References

Abbo, S., Saranga, B.Y., Peleg, B.Z., Kerem, B.Z., Lev-Yadun, B.S. and Gopher, B.A., 2009. Reconsidering domestication of legumes versus cereals in the ancient Near East. *The Quarterly Review of Biology* 84(1): 29–50.

Archer, S. and Pyke, D.A., 1991. Plant-animal interactions affecting plant establishment and persistence on revegetated rangeland. *Journal of Range Management*, 44(6), pp. 558–565.

Bagchi, R., Gallery, R.E., Gripenberg, S., Gurr, S.J., Narayan, L., Addis, C.E., Freckleton, R.P. and Lewis, O.T., 2014. Pathogens and insect herbivores drive rainforest plant diversity and composition. *Nature* 506(7486): 85.

Barton L.V., 1965. Dormancy of seeds imposed by the seed coat. *Encyclopedia of Plant Physiology* 15: 727–745.

Baskin, C.C., 2003. Breaking physical dormancy in seeds – focusing on the lens. *New Phytologist*, 158(2): 229–232.

Baskin, C.C. and Baskin, J.M., 1998. *Seeds-Ecology, Biogeography, and Evolution of Dormancy and Germination*. Academic, San Diego.

Baskin, C.C. and Baskin, J.M., 2014. *Seeds: Ecology, Biogeography, and Evolution of Dormancy and Germination* (2nd edn). Elsevier/Academic, San Diego, CA.

Baskin, J.M. and Baskin, C.C., 1989. Physiology of dormancy and germination in relation to seed bank ecology. In M.A. Lock, V.L. Parker and R.L. Simpson (eds) *Ecology of Soil Seed Banks*. Academic Press, INC, San Diego, CA, pp. 53–66.

Baskin, J.M. and Baskin, C.C., 2004. A classification system for seed dormancy. *Seed Science Research* 14(1): 1–16.

Benítez-Malvido, J., Tapia, E., Suazo, I., Villaseñor, E. and Alvarado, J., 2003. Germination and seed damage in tropical dry forest plants ingested by iguanas. *Journal of Herpetology* 37(2): 301–308.

Benvenuti, S., Macchia, M. and Miele, S., 2001. Light, temperature and burial depth effects on Rumex obtusifolius seed germination and emergence. *Weed Research* 41(2): 177–186.

Bewley, J.D. and Black, M., 2012. *Physiology and Biochemistry of Seeds in Relation to Germination: Volume 2: Viability, Dormancy, and Environmental Control*. Springer Science & Business Media, Berlin.

Bewley, J.D., Bradford, K. and Hilhorst, H., 2012. *Seeds: Physiology of Development, Germination and Dormancy*. Springer Science & Business Media, Berlin.

Bhatt, A., Carón, M.M., Verheyen, K., Elsarrag, E. and Alhorr, Y., 2016. Germination and seedling performance of five native legumes of the Arabian Desert. *Flora-Morphology, Distribution, Functional Ecology of Plants* 220: 125–133.

Blaney, C.S. and Kotanen, P.M., 2001. Effects of fungal pathogens on seeds of native and exotic plants: a test using congeneric pairs. *Journal of Applied Ecology* 38: 1104–1113.

Bouman, F. and Boesewinkel, F.D., 2017. The seed: structure and function. In J. Kigel (ed.) *Seed Development and Germination*. Routledge, London, pp. 1–24.

Bradstock, R.A. and Auld, T.D., 1995. Soil temperature during experimental bushfires in relation to fire intensity: consequences for legume germination and fire management in south-eastern Australia. *Journal of Applied Ecology* 32: 76–84.

Brewer, S.W. and Rejmanek, M., 1999. Small rodents as significant dispersers of tree seeds in a Neotropical forest. *Journal of Vegetation Science* 10: 165–174.

Brown, J.H., Reichman, O.J. and Davidson, D.W., 1979. Granivory in desert ecosystems. *Annual Review of Ecology and Systematics* 10(1): 201–227.

Campos, C.M. and Ojeda, R.A., 1997. Dispersal and germination of *Prosopis flexuosa* (Fabaceae) seeds by desert mammals in Argentina. *Journal of Arid Environments* 35(4): 707–714.

Cavanagh, T., 1987. Germination of hard-seeded species (Order Fabales). In P. Langkap (ed.) *Germination of Australian Native Plant Seed*. AMIRA, Melbourne/Sydney, pp. 58–70.

Chanthorn, W., Wiegand, T., Getzin, S., Brockelman, W.Y. and Nathalang, A., 2018. Spatial patterns of local species richness reveal importance of frugivores for tropical forest diversity. *Journal of Ecology* 106(3): 925–935.

Cho, H.K., Miyamoto, T., Takahashi, K., Hong, S.G. and Kim, J.J., 2007. Damage to *Abies koreana* seeds by soil-borne fungi on Mount Halla, Korea. *Canadian Journal of Forest Research* 37: 371–382.

Choudhury, B.I., Khan, M.L., Arunachalam, A. and Arunachalam, K., 2007. *Gymnocladus assamicus* Kanjilal ex PC Kanjilal fruit—a soap substitute. *Natural Product Radiance* 6: 427–429.

Choudhury, B.I., Khan, M.L. and Das, A.K., 2009. Seed dormancy and germination in *Gymnocladus assamicus*: An endemic legume tree from Northeast India. *Seed Science and Technology* 37(3): 582–588.

Choudhury, B.I., Khan, M.L. and Dayanandan, S., 2014. Functional androdioecy in critically endangered *Gymnocladus assamicus* (Leguminosae) in the eastern Himalayan region of northeast India. *PloS one* 9(2): e87287.

Clements, J., Jones, P.G. and Gilbert, N.H. 1977. Effect of seed treatments on germination in Acacia. *Australian Journal of Botany* 25: 269–276.

Cochrane, A., Yates, C.J., Hoyle, G.L. and Nicotra, A.B., 2015. Will among-population variation in seed traits improve the chance of species persistence under climate change? *Global Ecology and Biogeography* 24(1): 12–24.

Crist, T.O. and Friese, C.F., 1993. The impact of fungi on soil seeds: implications for plants and granivores in a semiarid shrub-steppe. *Ecology* 74: 2231–2239.

Dalling, J.W., Swaine, M.D. and Garwood, N.C., 1995. Effect of soil depth on seedling emergence in tropical soil seed-bank investigations. *Functional Ecology*: 119–121.

Danthu, P. M., Roussel, M., Dia, M. and Sarr, A., 1992. Effect of different pretreatments on the germination of *Acacia senegal* seeds. *Seed Science and Technology* 20: 111–117.

Dea, I.C. and Morrison, A., 1975. Chemistry and interactions of seed galactomannans. In R.S. Tipson and D. Horton (eds) *Advances in Carbohydrate Chemistry and Biochemistry* 31: 241–312.

Debeaujon, I., Léon-Kloosterziel, K.M. and Koornneef, M., 2000. Influence of the testa on seed dormancy, germination, and longevity in *Arabidopsis*. *Plant Physiology*, 122(2): 403–414.

Dinoor, A. and Eshed, N., 1984. The role and importance of pathogens in natural plant communities. *Annual Review of Phytopathology* 22: 443–466.

Domsch, K.H., Gams, W. and Anderson, T.H., 1993. *Compendium of Soil Fungi*, vol. 1. IHW-Verlag, Eching, Germany.

Egley, G.H., 1989. Water-impermeable seed coverings as barriers to germination. In R.B. Taylorson (ed.) *Recent Advances in the Development and Germination of Seeds*. Plenum Press, New York, pp. 207–233.

Fenner, M., 1983. Relationships between seed weight, ash content and seedling growth in twenty-four species of Compositae. *New Phytologist*, 95(4): 697–706.

Finch-Savage, G., Leubner-Metzger, G., 2006. Seed dormancy and the control of germination. *New Phytologist*: 171.

Footitt, S., Douterelo-Soler, I., Clay, H. and Finch-Savage, W.E., 2011. Dormancy cycling in *Arabidopsis* seeds is controlled by seasonally distinct hormone-signaling pathways. *Proceedings of the National Academy of Sciences* 108(50): 20236–20241.

Ganeshaiah, K.N. and Shaanker, R.U., 1992. Frequency distribution of seed number per fruit in plants: A consequence of the self-organizing process? *Current Science*: 359–365.

Ganeshaiah, K.N., Shaanker, R.U. and Shivashankar, G., 1986. Stigmatic inhibition of pollen grain germination – its implication for frequency distribution of seed number in pods of *Leucaena leucocephala* (Lam) de Wit. *Oecologia*, 70(4): 568–572.

Gardener, C.J., McIvor, J.G. and Jansen, A., 1993. Passage of legume and grass seeds through the digestive tract of cattle and their survival in faeces. *Journal of Applied Ecology*: 63–74.

Gong, X., Bassel, G.W., Wang, A., Greenwood, J.S. and Bewley, J.D., 2005. The emergence of embryos from hard seeds is related to the structure of the cell walls of the micropylar endosperm, and not to endo-β-mannanase activity. *Annals of Botany* 96(7): 1165–1173.

Graven, P., DeKoster, C. G., Boon, J. J. and Bouman, F., 1997. Functional aspects of mature seed coat of the Cannaceae. *Plant Systematics and Evolution* 205: 223–240.

Guariguata, M.R., 2000. Seed and seedling ecology of tree species in neotropical secondary forests: management implications. *Ecological Applications* 10(1): 145–154.

Gupta, S., Kaliamoorthy, S., Mao, A.A. and Soneswar, S., 2013. In-vitro regeneration of *Gymnocladus assamicus* Kanjilal ex PC Kanjilal, an ethnomedicinal critically endangered and endemic tree species from Northeast India. *International Journal of Medicinal and Aromatic Plants* 3(2): 204–213.

Hamrick, J.L. and Lee, J.M., 1987. Effect of soil surface topography and litter cover on the germination, survival, and growth of musk thistle (*Cardus nutans*). *American Journal of Botany* 74(3): 451–457.

Harper, J.L., Lovell, P.H. and Moore, K.G., 1970. The shapes and sizes of seeds. *Annual Review of Ecology and Systematics* 1(1): 327–356.

Harper, K.T., 1979. Some reproductive and life history characteristics of rare plants and implications of management. *Great Basin Naturalist Memoirs* 3: 129137.

Herranz, J.M., Ferrandis, P. and Martínez-Sánchez, J.J., 1998. Influence of heat on seed germination of seven Mediterranean Leguminosae species. *Plant Ecology* 136(1): 95–103.

Herranz, J.M., Ferrandis, P. and Martínez-Duro, E., 2010. Seed germination ecology of the threatened endemic Iberian *Delphinium fissum* subsp. *sordidum* (Ranunculaceae). *Plant Ecology* 211(1): 89–106.

Herrera, C.M., Jordano, P., Lopez-Soria, L. and Amat, J.A., 1994. Recruitment of a mast-fruiting, bird-dispersed tree: bridging frugivore activity and seedling establishment. *Ecological Monographs* 64(3): 315–344.

Howe, H.F. and Smallwood, J., 1982. Ecology of seed dispersal. *Annual Review of Ecology and Systematics* 13: 201–228.

Hughes, L., Dunlop, M., French, K., Leishman, M.R., Rice, B., Rodgerson, L. and Westoby, M., 1994. Predicting dispersal spectra: a minimal set of hypotheses based on plant attributes. *Journal of Ecology* 82: 933–950.

Hutchinson, G.E., 1951. Copepodology for the Ornithologist. *Ecology* 32: 571–577.

Igea, J., Miller, E.F., Papadopulos, A.S. and Tanentzap, A.J., 2017. Seed size and its rate of evolution correlate with species diversification across angiosperms. *PLoS biology* 15(7): p.e2002792.

Jacobsen, J.V. and Pressman, E., 1979. A structural study of germination in celery (*Apium graveolens* L.) seed with emphasis on endosperm breakdown. *Planta* 144(3): 241–248.

Jacquemyn, H., Brys, R. and Hermy, M., 2001. Within and between plant variation in seed number, seed mass and germinability of *Primula elatior*: effect of population size. *Plant Biology* 3(5): 561–568.

Jansen, P.A., Visser, M.D., Joseph Wright, S., Rutten, G. and Muller-Landau, H.C., 2014. Negative density dependence of seed dispersal and seedling recruitment in a Neotropical palm. *Ecology Letters* 17(9): 1111–1120.

Janzen, D.H., 1969. Seed-eaters versus seed size, number, dispersal, and toxicity. *Evolution* 23: 1–27.

Janzen, D.H., 1970. Herbivores and the number of tree species in tropical rain forests. *American Naturalist* 104: 501–528.

Janzen, D.H., 1978. Seeding patterns of tropical trees. In P.B.Tomlinson and M.H. Zimmermann (eds) *Tropical Trees as Living Systems*. Cambridge University Press, Cambridge, pp. 83–128.

Janzen, D.H., Demment, M.W. and Robertson, J.B., 1985. How fast and why do germinating Guanacaste seeds (*Enterolobium cyclocarpum*) die inside cows and horses? *Biotropica*: 322–325.

Jusaitis, M., Polomka, L. and Sorensen, B., 2004. Habitat specificity, seed germination and experimental translocation of the endangered herb *Brachycome muelleri* (Asteraceae). *Biological Conservation* 116(2): 251–266.

Kelly, D., 1984. Seeds per fruit as a function of fruits per plant in 'depauperate' annuals and biennials. *New Phytologist* 96(1): 103–114.

Kelly, K.M., Van Staden, J. and Bell, W.E., 1992. Seed coat structure and dormancy. *Plant Growth Regulation* 11(3): 201–209.

Khan, M.L. and Shankar, U., 2001. Effect of seed weight, light regime and substratum microsite on germination and seedling growth of *Quercus semiserrata* Roxb. *Tropical Ecology* 42(1): 117–125.

Khan, M.L. and Tripathi, R.S., 1987. Ecology of forest tree of Meghalaya: seed germination and survival and growth of *Albizzia lebbeck* seedlings in nature. *Indian Journal of Forestry* 10(1): 38–43.

Khan, M.L., Cavers, P.B., Kane, M. and Thompson, K., 1997. Role of the pigmented seed coat of proso millet (*Panicum miliaceum* L.) in imbibition, germination and seed persistence. *Seed Science Research* 7(1): 1–26.

Khurana, E. and Singh, J.S., 2001. Ecology of seed and seedling growth for conservation and restoration of tropical dry forest: a review. *Environmental Conservation* 28(1): 39–52.

Kolář, J., Kučerová, A., Jakubec, P. and Vymazal, J., 2017. Seed bank of *Littorella uniflora* (L.) Asch. in the Czech Republic, Central Europe: does burial depth and sediment type influence seed germination? *Hydrobiologia* 794(1): 347–358.
Kwak, M.M., Velterop, O. and van Andel, J., 1998. Pollen and gene flow in fragmented habitats. *Applied Vegetation Science* 1(1): 37–54.
Larsen, E., 1964. Germination response of *Acacia* seeds to boiling. *Australian Journal of Forest Research* 1: 51–53.
Larson, J.E., Sheley, R.L., Hardegree, S.P., Doescher, P.S. and James, J.J., 2015. Seed and seedling traits affecting critical life stage transitions and recruitment outcomes in dryland grasses. *Journal of Applied Ecology* 52(1): 199–209.
Lee, T.D., 1988. Patterns of fruit and seed production. *Plant Reproductive Ecology: Patterns and Strategies.* Oxford University Press, New York, pp.179–202.
Leishman, M.R., Masters, G.J., Clarke, I.P. and Brown, V.K., 2000. Seed bank dynamics: the role of fungal pathogens and climate change. *Functional Ecology* 14: 293–299.
Limón, Á. and Peco, B., 2016. Germination and emergence of annual species and burial depth: implications for restoration ecology. *Acta oecologica* 71: 8–13.
MacKay, B.D., 1972. The measurement of viability. In E.H. Roberts (ed.) *Viability of Seeds.* Chapman & Hall, London, pp.172–208.
Maron, J.L. and Crone, E., 2006. Herbivory: effects on plant abundance, distribution and population growth. *Proceedings of the Royal Society B: Biological Sciences* 273(1601): 2575–2584.
Masamba C., 1994. Pre-sowing seed treatments on four African *Acacia* species: appropriate technology for use in forestry for rural development. *Forest Ecology and Management* 64: 105–109.
McCleary, B.V. and Matheson, N.K., 1974. α-D-Galactosidase activity and galactomannan and galactosylsucrose oligosaccharide depletion in germinating legume seeds. *Phytochemistry*, 13(9), pp. 1747–1757.
Mello, M.A.R., Rodrigues, F.A., Costa, L.D.F., Kissling, W.D., Şekercioğlu, Ç.H., Marquitti, F.M.D. and Kalko, E.K.V., 2015. Keystone species in seed dispersal networks are mainly determined by dietary specialization. *Oikos* 124(8): 1031–1039.
Menon, S., Choudhury, B.I., Khan, M.L. and Peterson, A.T., 2010. Ecological niche modeling and local knowledge predict new populations of Gymnocladus assamicus a critically endangered tree species. *Endangered Species Research* 11(2): 175–181.
Michaels, H.J., Benner, B., Hartgerink, A.P., Lee, T.D., Rice, S., Willson, M.F. and Bertin, R.I., 1988. Seed size variation: magnitude, distribution, and ecological correlates. *Evolutionary Ecology* 2(2): 157–166.
Midgley, J.J. and Bond, W.J., 2001. A synthesis of the demography of African acacias. *Journal of Tropical Ecology* 17: 871–886.
Miller, M.F. and Coe, M., 1993. Is it advantageous for Acacia seeds to be eaten by ungulates? *Oikos*: 66(2) 364–368.
Mishra, Y., Rawat, R., Rana, P.K., Sonkar, M.K. and Mohammad, N., 2014. Effect of seed mass on emergence and seedling development in *Pterocarpus marsupium* Roxb. *Journal of Forestry Research* 25(2): 415–418.
Mohamed-Yasseen, Y., Barringer, S.A., Splittstoesser, W.E. and Costanza, S., 1994. The role of seed coats in seed viability. *The Botanical Review* 60(4): 426–439.
Moles, A.T., Ackerly, D.D., Webb, C.O., Tweddle, J.C., Dickie, J.B. and Westoby, M., 2005. A brief history of seed size. *Science* 307(5709): 576–580.
Nambara, E. and Nonogaki, H., 2012. Seed biology in the 21st century: perspectives and new directions. *Plant and Cell Physiology* 53(1).

Nelson, D.C., Flematti, G.R., Ghisalberti, E.L., Dixon, K.W. and Smith, S.M., 2012. Regulation of seed germination and seedling growth by chemical signals from burning vegetation. *Annual Review of Plant Biology*, 63.

Nonogaki, H., 2017. Seed biology updates–highlights and new discoveries in seed dormancy and germination research. *Frontiers in Plant Science* 8: 524.

Ofori, D.A., Newton, A.C., Leakey, R.R.B. and Grace, J., 1996. Vegetative propagation of *Milicia excelsa* by leafy stem cuttings: effects of auxin concentration, leaf area and rooting medium. *Forest Ecology and Management* 84(1–3): 39–48.

Patane, C. and Bradford, K.J., 1993. Seed dormancy in two Medicago species typical of natural pastures of Ragusa Plateau. *Rivista di Agronomia* 27: 412–418.

Paul-Victor, C. and Turnbull, L.A., 2009. The effect of growth conditions on the seed size/number trade-off. *PLoS One* 4(9): e6917.

Peco, B., Lopez-Merino, L. and Alvir, M., 2006. Survival and germination of Mediterranean grassland species after simulated sheep ingestion: ecological correlates with seed traits. *Acta Oecologica* 30(2): 269–275.

Poisot, T., Bever, J.D., Nemri, A., Thrall, P.H. and Hochberg, M.E., 2011. A conceptual framework for the evolution of ecological specialisation. *Ecology Letters* 14(9): 841–851.

Primack, R.B., 1978. Regulation of seed yield in *Plantago. The Journal of Ecology*: 835–847.

Primack, R.B., 1987. Relationships among flowers, fruits, and seeds. *Annual Review of Ecology and Systematics* 18(1): 409–430.

Quinlivan, B.J., 1971. Seed-coat impermeability in legumes. *Journal of the Australian Institute for Agricultural Science* 37: 283–295.

Radchuk, V. and Borisjuk, L., 2014. Physical, metabolic and developmental functions of the seed coat. *Frontiers in Plant Science* 5: 510.

Ratnamhin, A., Elliott, S. and Wangpakapattanawong, P., 2011. Vegetative propagation of rare tree species for forest restoration. *Chiang Mai Journal of Science*, 38: 306–310.

Reid, J.S.G., 1971. Reserve carbohydrate metabolism in germinating seeds of *Trigonella foenum-graecum* L. (Leguminosae). *Planta* 100: 131–142.

Reid, J.S.G. and Bewley, J.D., 1979. A dual role for the endosperm and its galactomannan reserves in the germinative physiology of fenugreek (*Trigonella foenum-graecum* L.), an endospermic leguminous seed. *Planta* 147: 145–150.

Roberts, E.H., 1973. Predicting the storage life of seeds. *Seed Science and Technology* 1: 499–514.

Roosmalen, M.G.M. van., 1985. Habitat preferences, diet, feeding behavior and social organization of the black spider monkey (*Ateles paniscus* Linnaeus 1758) in Suriname. *Acta Amazonica* 15(3–4): 238.

Russi, L., 1993. Preliminary observations on seed coat hardseededness in six pasture legumes. Developing annual pasture legumes for Mediterranean farming systems. In: S. Christiansen, L. Materon, M. Falcinelli and P. Cocks (eds) *Introducing Ley Farming to the Mediterranean Basin, ICARDA*: 289–293.

Salisbury, E.J., 1942. *The Reproductive Capacity of Plants*. G. Bell and Sons Ltd, London.

Shaanker, R.U., Ganeshaiah, K.N.A. and Bawa, K.S., 1988. Parent-offspring conflict, sibling rivalry, and brood size patterns in plants. *Annual Review of Ecology and Systematics* 19(1): 177–205.

Smýkal, P., Vernoud, V., Blair, M.W., Soukup, A. and Thompson, R.D., 2014. The role of the testa during development and in establishment of dormancy of the legume seed. *Frontiers in Plant Science* 5: 351.

Spear, E.R., Coley, P.D. and Kursar, T.A., 2014. Do pathogens limit the distributions of tropical trees across a rainfall gradient? *Journal of Ecology* 103(1): 165–174.

Stein, W.I., Slabaugh, P.E. and Plummer, A.P. 1974. *Seed of Woody Plants in the United States*, Agriculture Handbook No. 450. USDA Forest Service, Washington D.C., pp. 300–320.

Sy, A., Grouzis, M. and Danthu, P., 2001. Seed germination of seven Sahelian legume species. *Journal of Arid Environments* 49(4): 875–882.

Teketay, D., 1997. Germination ecology of *Acacia negrii*, an endemic multipurpose tree from Ethiopia. *Tropical Ecology* 38(1): 39–46.

Thapliyal, R.C., Rawat, M.M.S., Ramachandra, N.G. and Aswal, S.S., 1998. Pretreatment and conditions for testing germination of seeds of some common Indian *Acacia* species. *Seed Science and Technology* 26: 525–529.

Tran, V.N. and Cavanagh, A.K., 1984. Structural aspects of dormancy. In D.R. Murray (ed.) *Seed Physiology, vol. 2. Germination and Reserve Mobilization*. Academic Press, Sydney, pp. 1–44.

Traveset, A., Robertson, A.W. and Rodríguez-Pérez, J., 2007. A review on the role of endozoochory in seed germination. In A.J. Dennis, E.W. Schuff, R.J. Green and D.A. Westcott (eds) *Seed Dispersal: Theory and its Application in a Changing World*, pp.78–103.

Travlos, I.S., Economoua, G. and Karamanosa, A.I., 2007. Germination and emergence of the hard seed coated *Tylosema esculentum* (Burch) A. Schreib in response to different pre-sowing seed treatments. *Journal of Arid Environments* 68(3): 501–507.

Tybrik, K. (1991) *Regeneration of Woody Legumes in Sahel*. AAU Report 27, Aarhus University Press, Aarhus.

Vander Wall, S.B. and Beck, M.J., 2012. A comparison of frugivory and scatter-hoarding seed-dispersal syndromes. *The Botanical Review* 78(1): 10–31.

van Ulft, L.H., 2004. Regeneration in natural and logged tropical rain forest: Modelling seed dispersal and regeneration of tropical trees in Guyana. PhD thesis, Utrecht University, Utrecht, the Netherlands. Tropenbos-Guyana Series 12. Tropenbos-Guyana Programme, Georgetown, Guyana.

Venier, P., García, C.C., Cabido, M. and Funes, G., 2012. Survival and germination of three hard-seeded *Acacia* species after simulated cattle ingestion: the importance of the seed coat structure. *South African Journal of Botany* 79: 19–24.

Wadl, P.A., Rinehart, T.A., Dattilo, A.J., Pistrang, M., Vito, L.M., Milstead, R. and Trigiano, R.N., 2014. Propagation for the conservation of *Pityopsis ruthii*, an endangered species from the southeastern United States. *HortScience* 49(2): 194–200.

Walters, C., Berjak, P., Pammenter, N., Kennedy, K. and Raven, P., 2013. Preservation of recalcitrant seeds. *Science* 339(6122): 915–916.

Watanabe, T., 2002. *Pictorial Atlas of Soil and Seed Fungi: Morphologies of Cultured Fungi and Key to Species* (2nd edn). CRC Press LLC, Boca Raton, FL.

Weber, H., Borisjuk, L. and Wobus, U., 1996. Controlling seed development and seed size in *Vicia faba*: a role for seed coat-associated invertases and carbohydrate state. *The Plant Journal* 10(5): 823–834.

Williams, P.R., Congdon, R.A., Grice, A.C. and Clarke, P.J., 2004. Soil temperature and depth of legume germination during early and late dry season fires in a tropical eucalypt savanna of north-eastern Australia. *Austral Ecology* 29: 258–263.

Willson, M.F., 1993. Dispersal mode, seed shadows, and colonization patterns. *Vegetatio* 107–108: 261–280.

Zar, J.H., 1974. *Biostatistical Analysis*. Prentice-Hall, Englewood Cliffs, NJ.

5 Seedling Dynamics and Ecology of *Gymnocladus assamicus*

5.1 Introduction

The seedling phase is a very sensitive stage in plant life cycle and several ecological factors closely associated with seedling establishment and growth affect regeneration of any forest tree species. Seedlings represent an intermediate stage of regeneration between seed and mature tree. For successful establishment, seedlings go through different stages and often face both biotic and abiotic challenges. Seedling performance depends on many factors including microsite, dispersal and herbivory. The seedling phase in the life of a tree is very important and shapes the future composition of the species in forests. According to Janzen–Connell hypothesis, seedlings perform better when they are far from the parent (Janzen, 1970; Connell, 1971). In other words, specialist natural enemies (e.g. insect pests, herbivores and pathogens) reduce the chances of survival rates of conspecific seeds and seedlings regenerated near to mother plants or in areas of high conspecific density and thereby maintain high diversity in plant communities in a natural ecosystem (Janzen, 1970; Connell, 1971; Latorre *et al.*, 2013).

Population dynamics of a plant species can be explained by various demographic features such as recruitment, mortality and growth rates (Watkinson, 1996). Population size of a species largely depends on birth, survival and death rates and varies considerably within species (Roff, 1992). For a species to maintain a viable population structure, balance among these variables has been found to be important (Kohyama and Hara, 1989). Several factors such as flowering, fruiting, seed maturation, dispersal and seedling establishment plays critical role in the life cycle of a species. For example, mature and hermaphrodite soap pod trees (*G. assamicus*) are able to produce a sufficient amount of seeds every year (Choudhury,

2008); however, they are unable to maintain a viable population structure, indicating potential constraints to natural regeneration in the natural habitat.

Seed germination initiates the process of seedling establishment and is complete when the seedlings attain photosynthetic competence and no longer depend on seed reserves for their survival. The establishment phase gives the biggest demographic squeeze to the seedling population (Whitmore, 1996) and it was found that about 50% of seedling mortality took place in the first year in two forest tree species (*Aspidosperma fendleri* and *Richena grandis*) studied in Venezuelan cloud forest (Flores, 1992). Availability of light in the form of photosynthetic active radiation (PAR) is an important determinant of seedling performance on forest floors. In general, PAR at ground level beneath a continuous tree canopy is around 1–3% in full sun conditions (Chazdon and Fetcher, 1984; Pearcy, 1987; Osunkjoya *et al.*, 1992). Light irradiance and spectral quality significantly contribute to tree seedling development and establishment in forest ecosystems (Lee *et al.*, 1996).

Tree seedling survival also depends on canopy characteristics such as canopy age, height, cover and composition (Laughlin and Clarkson, 2018). Pioneer species are generally able to tolerate harsh environmental conditions such as drought and winter frost and grow rapidly to establish (Lusk and Laughlin, 2017). Non-pioneer species (or climax species), on the other hand, can germinate under forest shade and seedlings can survive and establish successfully in such environments (Swaine and Whitmore, 1988). Seedling mortality is a common phenomenon in both natural forest and silvicultural practices. High mortality at seedling stage can strongly filter the composition of the species of the seedling bank which accumulate over multiple years and cannot emerge from the understory due to canopy cover (Whitmore, 1984; Brokaw, 1985). Such spacing mechanisms, known as negative density-dependent mortality, promote species' coexistence by preventing the same species from becoming too locally abundant (Metz *et al.*, 2010). Moreover, conspecific seedling densities in the vicinity of the mother tree of the same species may also lead to greater chances of mortality due to specialist herbivores or pathogens (Packer and Clay, 2000). At the same time, high seedling mortality may act as a strong selective filter on seed and seedling traits as well (Kitajima and Fenner, 2000).

Researchers have identified several components of regeneration in a forest ecosystem of which seed dispersal, method of seed dispersal, frequency and spatial as well as size distribution of canopy gaps, and gap characteristics are critical (Fredericksen and Licona, 2000; Dalling *et al.*, 2002; Grau, 2002; Kukkonen *et al.*, 2008). Success of regeneration of trees in a forest depends upon the response of naturally growing and transplanted seedlings to the prevailing microenvironment (Whitmore, 1975). It was found that seedlings in the below-canopy seedling bank grow very slowly in height but then quickly commence rapid height growth after a canopy gap develops above them, a process known as seedling release (Whitmore, 1996). In temperate forests, multiple releases take place where seedlings very commonly grow to tree size after a series of releases, pulses of rapid growth followed by stasis, in successive gaps forming at the same place over many years (Runkle and Yetter, 1987).

A seedling is exposed to various biotic and abiotic factors and interactive influences of both ultimately determine the survival and growth at a particular site (Augspurger, 1983; Johnson *et al.*, 2017). Major factors that affect seedling performance are light or PAR (Mori *et al.*, 1990; Sasaki and Mori, 1981; Turnbull, 1991; Coomes and Grubb, 2000), sunflecks (Chazdon, 1988; Pearcy, 1990), mineral nutrients (Stewart *et al.*, 1990), water availability (Ashton, 1990; Whitmore, 1990), pathogen (Mueller-Dombois *et al.*, 1983) and burning (Williams *et al.*, 2005). Ectomycorrhizae play an important role in seedling growth and establishment of dipterocarp forest (Smits, 1983). Abiotic stresses may also indirectly affect seedlings by lowering their tolerance to biotic agents such as herbivores and pathogens (Augspurger and Kelly, 1984; Fenner, 1987; Johnson *et al.*, 2017). Legume tree seedlings mostly develop mycorrhiza and root nodulation early in the seedling stage, helping them to establish quickly (Hogberg, 1986; Michelsen, 1992; Birhane *et al.*, 2015).

Several ecological factors closely associated with seedling establishment and growth affect regeneration of any forest tree species. A few major factors are regeneration strategy (Ng, 1978; Swaine and Whitmore, 1988; Whitmore, 1996), seed size (Foster, 1986; Khan, 2004; Moles and Westoby, 2004), germination (Garwood, 1983), life form (Hladik and Miquel, 1990) dispersal mode (Ng, 1978; Hladik and Miquel, 1990) and habitat (Duke, 1970; de Vogel, 1980). Microsite is an important parameter for seedling establishment and largely influences the growth, survival and establishment of a species, especially in a forest ecosystem (Frehner, 2002; Hunziker and Brang, 2005; Bertacchi *et al.*, 2016; Pernot *et al.*, 2019). For example, some species can grow in low-light environments and have less specific microsite requirements than other species (Hunziker and Brang, 2005).

Seedling establishment is also largely impacted by exposure to various external agents such as pathogen vulnerability, herbivory and physical damage. High seedling density was found to be positively correlated with disease levels (Augspurger and Kelly, 1984). Herbivores have the ability to damage seedlings at different degrees and can cause partial or complete damage to leaves and apical meristems (Clark and Clark, 1984). Herbivore damage can potentially affect seedling survival and longevity. It was found that a natural seedling population with higher density has earlier onset of herbivory than for transplanted seedlings to sites with no conspecific seedlings within 10 m distance (Clark and Clark, 1984). Direct and indirect influence of insect herbivory have been documented for many plant species and communities, thereby impacting plant community composition by altering the recruitment, mortality or individual growth rates (Maron and Crone, 2006; Kim and Underwood, 2015; Myers and Sarfraz, 2017). Physical damage of seedlings is a major cause of mortality and can directly influence the process of natural regeneration in a forest ecosystem. Physical damage accounted for 82% loss of artificial seedlings in a Costa Rican rain forest (Clark and Clark, 1989). Physical damage by litterfall is another potential cause of seedling mortality and may cause up to 35% annually in fragmented forests, which is upto 24% in continuous forest (Scariot, 2000).

Legume taxa represents a significant component of nearly all terrestrial biomes, on all continents (except Antarctica) and are ecologically very successful in terms of dominance and productivity. Members of the family show an enormous amount of diversity in their life forms and exhibit a tremendous range of adaptations. These traits are expressed through remarkable levels of adaptive variation in morphological characteristics such as growth form, canopy architecture, root architecture, leaf morphology, pod and seed structure, and in physiological features such as phenological controls, solar tracking, water relations, hard seed coats, nitrogen fixation and mycorrhizal associations (Rundel, 1989; Yahara *et al.*, 2013). Legumes can also employ stem photosynthesis, which in certain cases may amount to about one third of the leaf tissue photosynthesis, with high water use efficiency of the fixed carbon. Phyllodes which are often developed as a sclerophyllous substitute of leaf tissue also have high water use efficiency (Rundel, 1989). Legumes have efficient water economy by different adaptations such as nyctinastic closure of leaves and the phreatophytic habit in many desert legumes, as in *Prosopis glandulosa* (Nilsen *et al.*, 1981). They also have adaptive mechanisms related to nitrogen fixation through the development of nitrogen-fixing root nodules in soils with limited nitrogen availability and formation of mycorrhizal associations which enhances phosphorus uptake more efficiently (Arianoutsou and Thanos, 1996). Legumes adopt various dispersal modes (endozoochory, autochory, etc.) with almost ubiquitous presence of hard seed coat that ensures maximum seed survival (when fruits are consumed by animals) that contributes to the formation of permanent and, long-lived soil seed banks (Arianoutsou and Thanos, 1996). Such species have several advantages and are variously used in reclamation of abandoned fields (Li *et al.*, 2009), development of grazing systems (Dear and Virgona, 1996) and as fodder and bee forage plants (Snook, 1982).

As *G. assamicus* is a critically endangered legume species, understanding seedling ecology and ecophysiology is crucial for developing strategies to manage forest ecosystems in general and any species of interest in particular. However, there is insufficient information on the seedling behaviour under different growing conditions. Seedling survival and establishment is site- and species-specific manner (Ashton, 1998) and studying different aspects of seedling ecophysiology is immensely helpful in saving the species from risks of extinction. In this chapter, seedling dynamics of *G. assamicus* has been explained starting from seed germination to seedling establishment both in situ and ex situ conditions in two different ecological environments. For in situ studies, available seedlings were monitored in the field (Dirang) where seedlings grow naturally under forest cover. Ex situ experiments were conducted in the greenhouses at Dirang Orchid Orchard and Department of Forestry, North Eastern Regional Institute of Science and Technology (NERIST), Nirjuli, Arunachal Pradesh, during 2005–2007. Dirang and NERIST sites fall under eastern Himalayan wet temperate forest (II-12B/CIa) and Assam alluvial plains semievergreen forest (2B/C1) respectively (Champion and Seth, 1968). The study also attempted to identify different functional morphological attributes of the seedlings, seedling survival and new recruitment, and growth performance under in situ and ex situ conditions. Performance of nursery-grown seedlings transplanted to their natural habitat was also studied.

5.2 Materials and Methods

Seedling ecology of *G. assamicus* was studied both in situ and ex situ conditions to identify different growth phases and challenges the seedlings face during their early stages of development. While the in situ study provides more insight on seedling mortality and growth rates in natural habitats, the ex situ study is helpful to develop effective silvicultural strategies for population restoration in suitable habitats.

5.2.1 In situ experiments

Natural regeneration of *G. assamicus* was found to be very poor, having seedling and sapling populations only at the Moishing site (27°18′ N latitude; 92°14′ E longitude; 2052 m asl) in the West Kameng District of Arunachal Pradesh (Fig. 1.12; Chapter 1). Therefore, in situ experiments were conducted only at Moishing and no replicates could be added to the analysis. Seedlings were located and identified in the field by carefully surveying the forest floor primarily in and around fruiting *G. assamicus* trees. Individual seedlings were permanently labelled with aluminium tags and different parameters related to survival and growth such as height, collar diameter and number of leaves per individual were recorded at six-month intervals for two consecutive years (2005–2007). Altogether, 54 seedlings were tagged in April 2005 and growth parameters were recorded in non-destructive ways because of the rarity of the species and to save the seedling population in its natural habitat. Height of individual seedlings was measured using measuring tape. Collar diameter of the seedlings was measured at ground level using digital calipers and the number of leaves was counted at the end of the rainy season before leaf fall was initiated. A seedling with a specific tag was considered dead if it was not encountered during the next census period. Growth rates in terms of percentage were calculated to assess seedling performance.

5.2.2 Ex situ experiments in the nursery

The ex situ experiments were conducted in two locations to evaluate the response of the seedlings in different ecological conditions. The first site was at Dirang (27°08′ N latitude; 92°4′ E longitude; 2052 m asl) in West Kameng District and the second site was at the NERIST Botanical Garden (27°07′ N latitude; 93°22′ E longitude; 110 m asl) located in Papum Pare District, Arunachal Pradesh. Reasons for selecting the NERIST site were its distinctive climatic conditions, native range of species, availability of a large botanical garden for field experiment and easy accessibility for regular monitoring (as we were stationed at NERIST). It is to be highlighted here that both the topography and climatic conditions differ dramatically between the two experimental sites (Table 5.1). Primary objectives of the ex situ trail was to study the seedling performance in wider geographical areas and develop strategies for reintroduction of the species, for example in roadside plantations or social forestry programmes.

Mature pods were harvested during the month of January based on phenological observations (Fig. 3.3, Chapter 3) and transported to the laboratory for sorting and subsequent experimental procedure. Two hundred seeds from randomly chosen pods were picked and divided equally for Dirang and NERIST nurseries. Our study demonstrated that more than 70% of seeds are viable within the first few weeks of harvesting (Choudhury *et al.*, 2009) and were ready for setting up germination. Polythene bags were filled with about 1 kg of farmyard soil and kept in the greenhouse under appropriate experimental conditions. One seed in each bag was sowed and watered regularly to maintain constant soil moisture. Seeds sown on the soil surface were found to be mostly infested by soil-borne fungi that damaged the seeds and prevented germination (Fig. 5.1). Regular watering was done to minimize water stress during the experimental period. Seedlings were raised under three light irradiance levels classified as low, medium and high by manipulating sunlight penetration in the greenhouse using shade cloth. In order to evaluate seedling performance in terms of survival and growth under different ecological conditions, growth parameters such as shoot elongation, collar diameter and leaf number of tagged seedlings were measured over a period of one year at three-month intervals (during November to October, 2006–2007). All growth parameters were recorded by non-destructive methods during the experiment. Final data for ex situ experiments were collected after harvesting the seedlings to calculate root and shoot biomass accumulation

Table 5.1. Topography and climatic conditions of different study sites (data recorded during 2005–2007).

	Natural	Nursery		Transplanted	
Parameter	Moishing	Dirang	NERIST	Dirang	NERIST
Mean altitude (m)	2,052	1,715	110	1,715	110
Max/Min air temperature (°C)	1–34	1–34	20–38	1–34	20–38
PAR (µmol s^{-1} m^{-2})	38–45	500–540	130–180	1,000–1,100	1,500–1,700
Annual precipitation (mm)	1,800–2,000	1,800–2,000	1,572–3,076	1,800–2,000	1,572–3,076
Average soil moisture (%)	30–35	35–40	35–40	17–20	15–20
Water-holding capacity (%)	60–65	60–65	60–70	45–50	40–45

Fig. 5.1. Seed damage by soil-borne fungi in nursery (photo: Baharul I. Choudhury).

pattern. Growth rates (seedling height, collar diameter and number of leaves) were calculated in terms of percentage growth over the study period.

5.2.3 Performance of reintroduced seedlings in natural habitats

To assess the performance of reintroduced seedlings, 50 seedlings (of about four months of age) grown in Dirang (Fig. 5.2) and NERIST (Fig. 5.3) nurseries were transplanted during April 2005 in each site respectively. Seedlings were tagged using permanent labels and numbered for future studies. Reintroduction was mainly carried out with community participation by distributing seedlings among the local people. Growth and survival of transplanted seedlings were monitored periodically.

Different microclimatic characteristics of each of the study plots are presented in Table 5.1. Geographical location and altitude of the study sites were recorded using handheld GPS (make: Garmin). Data on maximum and minimum air temperature and annual precipitation rates were collected from the State Meteorology Department and NERIST Meteorology Center during the study period. Soil moisture content was determined following the Keen's box method (Piper, 1944). Water-holding capacity was measured gravimetrically by incubating 10 gm of field moist soil sample in a hot-air oven at 105°C for 24 h. Quantum of light in terms of photosynthetic active radiation (PAR) was measured using a LICOR plant canopy analyser. Leaf area of the harvested nursery-grown seedlings was measured with the help of a digital leaf area meter (Systronics model 211) and biomass was calculated on dry weight basis by oven drying fresh plant materials at 100°C for 48 h. Statistical analyses were conducted using the software program STATISTICA (StatSoft). Analysis of variance (ANOVA) was performed to assess the significance levels between different treatment types. Regression analysis was performed to understand the interaction between seedling mortality and different environmental factors such as rainfall, minimum and maximum temperature.

Fig. 5.2. Seedlings grown in Dirang nursery ready for transplantation (photo: Baharul I. Choudhury).

Fig. 5.3. Seedlings grown in NERIST nursery ready for transplantation (photo: Baharul I. Choudhury).

5.3 Results and Discussion

5.3.1 Seedling functional morphology

Forest trees display an extraordinary diversity of seedling morphologies. Major variations occur in the form of arrangement, relative sizes and contrasting shapes of leaves and cotyledons as well as other structures such as scale leaves and pubescence. Leguminosae display two basic seedling types based on the consistency of the cotyledons, namely 'phyllolobees' which are leaf-like cotyledons and 'sarcolobees' with storage cotyledons (De Candolle, 1825). At the beginning, cotyledons in *G. assamicus* escape from the seed coat and are carried above the soil, remaining just above ground level (Fig. 5.4). The cotyledons also display rich amounts of food reserve and supply nutrient requirements at the early stages of seedling establishment. Based on early emerging trends, *G. assamicus* seedlings are categorized as phanerocotylar-hypogeal-reserve (PHR) under broad classification of phyllolobees type which is common in other members of the leguminosae family (Rodrigues, 2015). PHR-type seedlings are equipped with food reserves that can withstand a certain level of unpredictable climatic conditions and confer tolerance to survive in such situations (Menezes *et al.*, 2018). Menezes *et al.* studied seedling types across rainfall gradient and found higher numbers of species with PHR seedling

Fig. 5.4. PHR seedling of *G. assamicus* (photo: Baharul I. Choudhury).

types in sites with decreasing average rainfall. Such mechanisms are indicative of advantageous evolutionary strategies to provide insurance to the delicate seedling stage when they are likely to be exposed to unfavourable environmental conditions. On the other hand, based on seedlings with storage reserves in the hypocotyl or root, seedlings of *G. assamicus* may also be classified as fleshy cryptocotylar as the cotyledons are specialized haustorial organs and never emerge from the seed, having enough food reserves in the endosperm (Garwood, 1996).

5.3.2 Survival and growth of seedlings under in situ conditions

Seedling mortality is common in natural forests and has been observed in *G. assamicus* seedling site in Moishing. Seedling count data showed around 13% death in the first six months from April to October 2005 (rainy season) which went up to about 23% between October 2005 and April 2006. Overall, more than 42% of the tagged seedling deaths were recorded during the two-year study period (between April 2005 to October 2006). Several factors are responsible for seedling mortality in a forest ecosystem including herbivory by animals (Clark and Clark, 1984) and insect damage (Maron and Crone, 2006; Myers and Sarfraz, 2017). A few *G. assamicus* seedlings were also found to be physically damaged, most likely due to litterfall, treefall or due to anthropogenic activities such as logging or human movement. Physical damage of seedlings has also been reported in other studies and accounted for as high as 82% of seedling loss in an experimental plot (Clark and Clark, 1989).

Higher seedling mortality was recorded during the drier season compared to the rainy season (October 2005–April 2006). At least 10% more seedlings died in the first dry season of the study period when rainfall was at a minimum. Water deficit is a major cause of seedling death in natural ecosystems (Turner, 1990; Brown and Whitmore, 1992) and increased seedling mortality of *G. assamicus* during drier seasons may be directly associated with water deficiency during the period. Swaine *et al.* (1990) and Lieberman and Li (1992) also found that survival rates of the seedlings of tree species increased progressively during the wet season. Soil moisture plays an important role in seedling growth and has been studied by many researchers (McLeod and Murphy, 1977; Mueller-Dombois *et al.*, 1980; Quero *et al.*, 2006; Bolte *et al.*, 2016). Effects of soil moisture on mortality of seedlings are also reported for many species in different areas (Mueller-Dombois *et al.*, 1980; Rao and Singh, 1985; Khan *et al.*, 1986; Kumar *et al.*, 1994; Burslem, 1996; Swinfield *et al.*, 2012; Andrus *et al.*, 2018).

Survival and growth of seedlings are also influenced by temperature (Hunt and Halligan, 1981; Burslem *et al.*, 1995; Ter Heerdt *et al.*, 2017). Cold temperature is known to be a major cause of stress and damage to tree seedlings (Blennow and Lindkvist, 2000; Chang *et al.*, 2015). The western Arunachal area experiences extreme weather events marked by subzero temperatures between December and February along with a modest amount of snowfall every year. Seedlings of *G. assamicus* are generally exposed to chilling weather conditions during winter months and are vulnerable to death due to cold stress. This demonstrates that low temperature is an important determinant for seedling mortality (Blennow and Lindkvist, 2000; Sakai and Larcher, 2012).

Seedling growth performances among all the sites were always better between April and September, i.e. during the rainy season compared to the dry and cool season, i.e. October to March. Superior seedling growth rates in terms of height, collar diameter and leaf number may be due to favourable growing conditions and high soil moisture content during that period. Furthermore, better growth performances during the rainy season may be attained due to the increased availability of soil nutrients because of rapid decomposition of leaf litter (Devi, 2004). Differences in relative growth rates among species are caused by morphological parameters, while temporal changes are caused by physiological parameters (Antúnez, *et al.*, 2001). Moishing and Dirang areas experience subzero temperatures with intermittent snowfall in the drier months (December–January) which might also play a role in seedling death as demonstrated in other species (Bigras and Dumais 2005; Reeves *et al.*, 2016).

Activities of grazing animals dramatically increase in the area during dryer months and more animals (hill cattle, goats and yaks) were encountered during field visits. It is possible that such animals also damage *G. assamicus* seedlings in the field (Scariot, 2000; Sankey *et al.*, 2006). However, other causes such as litterfall or unknown reasons may also be responsible for seedling disappearance in the natural habitat (Scariot, 2000).

Higher seedling survival was recorded at distances further from the mother plant, indicated by the occurrence of sapling population at 9 to 12 m radial distance (Fig. 5.5). Higher seedling survival at greater distance could be most likely due to less competition beyond the parent tree and spatial differences in microenvironment such as leaf litter and soil nutrients along the different distances (Janzen, 1970; Connell, 1971).

Growth of seedlings in natural conditions showed an increasing trend in terms of shoot length, collar diameter and number of leaves over time. Growth rate of shoot length and collar diameter was found more during April–September than October–March. Similarly, the number of leaves increased during April–September and decreased during the dry season. Because of very slow growth rates of *G. assamicus*, no seedlings could have been found to attain sapling stage during the two-year study period.

5.3.3 Growth and survival of seedlings under ex situ conditions

G. assamicus seedling survival rates differed under low, moderate and high light irradiance levels in the nurseries. Maximum survival rate (94%) was documented under moderate light irradiance level during a one-year study period in NERIST nursery followed by low and high light irradiance levels (Fig. 5.6). In another study, intermediate light levels also showed optimal growth in cherrybark oak (*Quercus pagoda*) seedlings in field conditions (Gardiner and Hodges, 1998). Seedling survival

Fig. 5.5. *G. assamicus* sapling population at Dirang site near a seasonal stream (photo: Baharul I. Choudhury).

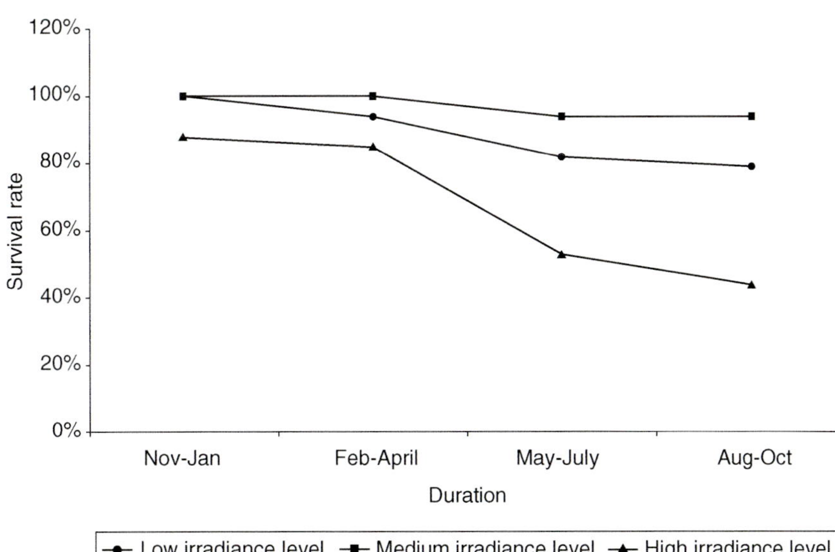

Fig. 5.6. *G. assamicus* seedling survival rates in NERIST nursery under low, medium and high light irradiance levels.

results showed significant variation (*p*-value = 0.04) between the irradiance levels. In Dirang nursery, maximum seedling death (18%) was recorded during the early stages of development within the first six months of growth. Highest seedling mortality at the very early stages of development is common in many forest species, even in the cotyledon stage (Purdie, 1977).

Seedlings performed extremely well under moderate light levels in which as high as an 81% increase in seedling height was recorded during November–January (Fig. 5.7). On the other hand, 27–63% and 6–35% increase in seedling height was recorded under low and high light irradiance levels respectively during different time periods (Fig. 5.7). Remarkable morphological variation in terms of seedling height was noticeable in the greenhouse under different light irradiance levels (Fig. 5.8). Canham *et al.* (1996) showed improved seedling performance of red maple (*Acer rubrum*) seedling under moderate light levels. Again, growth rates in terms of height varied significantly (*p*-value = 0.08) between the treatment types as estimated by ANOVA. Enhanced seedling growth for many forest species in gaps has been reported by Augspurger and Kelly (1984) and Pompa and Bongers (1988). Different authors have shown similar results in the case of *Shorea* species (Ashton 1990) and *Quercus semiserrata* seedlings (Khan and Shankar, 2001) in natural forest ecosystems. However, better growth and survival of seedlings of different species in sunny areas have also been reported by many researchers (Whitmore, 1975; Lambers and Poorter, 1992; Gómez-Aparicio *et al.*, 2008).

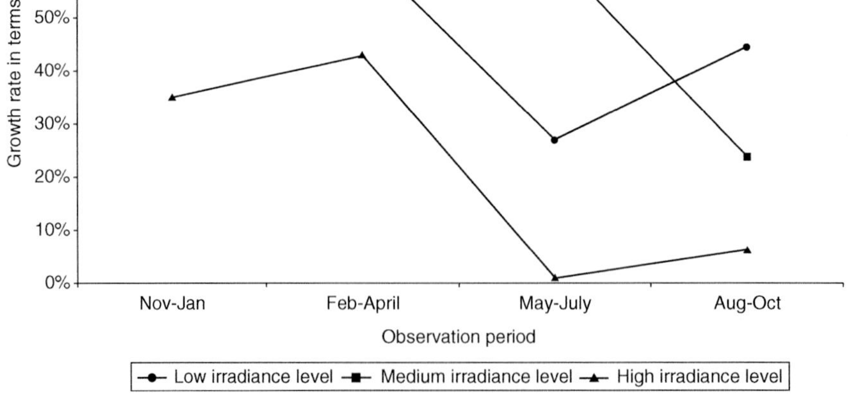

Fig. 5.7. *G. assamicus* seedling growth rates in terms of height under low, medium and high light irradiance levels.

Fig. 5.8. Morphological variation in seedling height under different light irradiance levels in the NERIST greenhouse (photo: Baharul I. Choudhury).

Higher shoot length enlongation (6 cm in one year) was recorded in Dirang compared to NERIST nursery (5 cm in one year). Average increase in shoot length was also found to be more in Dirang nursery compared to NERIST nursery. Seedling growth rate in terms of collar diameter also increased with time.

Highest increase in collar diameter recorded was 0.61 mm in Dirang nursery during April–September. Overall, 30–60% collar diameter increment was recorded under medium light irradiance followed by low (19–52%) and high (15–28%) irradiance levels (Fig. 5.9). Number of leaves also increased with increase in age and was always more during the rainy season than the drier season. Similar responses have also been reported in the case of tropical trees (Augspurger 1983), subtropical trees (Khan *et al.*, 1986) and *Dipterocarpus* seedlings (Sasaki and Mori, 1981).

Seedling mortality during the early stages of growth is very common in many forest tree species. A high proportion of seedlings may be attacked by

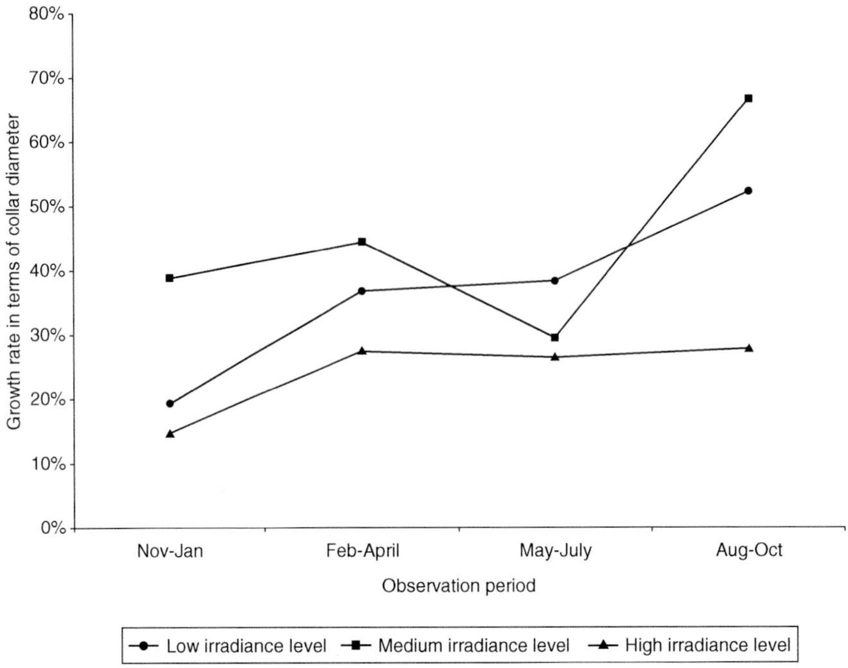

Fig. 5.9. *G. assamicus* seedling growth rates in terms of collar diameter under low, medium and high light irradiance levels.

phytopathogens at the early growth phase (Purdie, 1977; Steege, 1990). Cotyledon damage by insects and pathogens was noticeable in greenhouse and nursery conditions (Fig. 5.10). Litterfall, particularly branches and trunks, is an important cause of mortality in seedlings and juveniles of many tropical tree species (Clark and Clark, 1989, 1991). On ageing, seedlings initiate symbiotic relationships for better growth and survival. Temperate legume seedlings fix nitrogen within two weeks of germination (Halvorson *et al.*, 1991). Tropical *Shorea* seedlings had ectomicorrhizal infections within 20 days of germination (Alexander *et al.*, 1992). In the case of *G. assamicus*, seedling mortality decreased with increase in age in Dirang site, though nodule formation was not observed at the age of two years. However, seedling mortality rates showed the opposite trend in the NERIST site. Very poor seedling growth and survival of transplanted seedlings in NERIST botanical garden may be mainly due to spatial heterogeneity and major differences in microclimatic conditions to those of the native *G. assamicus* population in Dirang. Garwood (1996) reported similar results and found that growth of tree seedlings in particular environments and their ability to adapt to changes in those environments depended on the complex interaction of morphological and physiological attributes of each species.

Biomass accumulation pattern in root, shoot and leaf in one-year old seedlings showed remarkable differences under different treatment levels and between nurseries. Biomass accumulation patterns were higher in Dirang nursery compared to NERIST. Yet, seedlings grew taller in NERIST than in Dirang (Figs 5.2 and 5.3). Highly developed root system (Fig. 5.11) was observed which had higher biomass than the aboveground shoot part. Average root and shoot biomass of one-year old seedlings was 3.3 gm (±0.53) and 1.43 gm (±0.15) respectively in Dirang and 2.2 gm (±0.31) and 1.41 gm (±0.15) respectively in NERIST. However, the variation in root biomass between the two nurseries was insignificant. This study clearly demonstrated that the *G. assamicus* seedlings are light-dependent and their survival will depend largely on canopy openess. Seedling damage by insect pests was also noticeable in the greenhouse (Fig. 5.12) which reflects the potential of similar

Fig. 5.10. *G. assamicus* cotyledon damage by insects as recorded in the nursery (photo: Baharul I. Choudhury).

Fig. 5.11. Two-year old *G. assamicus* seedlings showing well-developed root system (photo: Baharul I. Choudhury).

Fig. 5.12. *G. assamicus* seedling damage by insect pests in the greenhouse. Noticeable foliar damage by caterpillar (highlighted in circle) was observed during the greenhouse experiment (photo: Baharul I. Choudhury).

consequences in field conditions. Pathogen and insect damage may result in as high as 34% leaf loss during seedling stages in forest species (Myster, 2002) and can damage both understory and canopy resources in natural ecosystems (Basset *et al.*, 1999).

5.3.4 Survival and growth of transplanted seedlings

There was almost similar growth rates among transplanted seedlings in Dirang and NERIST sites during the first six months of growth. However, significantly lower average shoot elongation was recorded in NERIST compared to Dirang site ($F=42.536$, $p<0.001$). Increase in shoot length of individual seedlings also differed significantly in each site (Dirang: $F=8.369$, $p<0.001$; NERIST: $F=13.704$, $p<0.001$). Furthermore, increase in collar diameter also varied significantly between the two sites ($F=7.769$, $p<0.01$). Seedling survival varied site-wise and was highest in Dirang nursery followed by Moishing and transplanted seedlings in Dirang. Brown and Whitmore (1992) found varied survival responses in naturally established seedling populations of three dipterocarp species to a range of gap size. Heterogeneous survival responses in the case of *G. assamicus* may be due to differences in micro-environmental conditions in those areas. Gap size and, conversely, the irradiance level, play a major role in the survival of tree seedlings (Brokaw, 1984) and this may be attributed to the differential survival response of *G. assamicus* seedlings in different sites.

Seedling growth in terms of shoot length among natural, transplanted and nursery conditions also showed remarkable variation. Highest rate of shoot elongation was found among seedlings in Dirang nursery followed by Moishing (field). Transplanted seedlings at Dirang (Fig. 5.13) performed much better both in terms of survival and growth than at NERIST. Overall, seedling survival in different sites also showed remarkable differences and maximum survival (54%) was recorded in nursery conditions at Dirang (Fig. 5.6). The least seedling survival was recorded in NERIST and only 28% of seedlings survived at the end of two years study period (Fig. 5.6).

Herbivory was one of the major causes of seedling mortality among all the study sites. Major damage of seedlings took place during the dry season (December–March) when domestic animals graze randomly in the forest areas (Fig. 5.14). This might have a potential negative impact on seedling survival

Fig. 5.13. Transplanted *G. assamicus* seedling at Dirang site (photo: Baharul I. Choudhury).

Fig. 5.14. Grazing animals under a *G. assamicus* tree at Dirang site (photo: Baharul I. Choudhury).

and growth in the natural habitat. Moreover, human activities such as harvesting fuel wood, non-timber forest produce and agricultural practices might also played a role in seedling damage during the drier season. Pathogen attack and insect infestation were also observed to cause mortality in *G. assamicus* seedlings. Insect herbivory can significantly increase seedling mortality, thereby affecting the performance of an individual species as well as the forest ecosystem (Lemoine *et al.*, 2017). Various studies suggested continuous decrease in seedling mortality with size and age of seedlings (De Steven, 1991; Tanouchi *et al.*, 1994).This is mainly due to their small size and soft and palatable tissue required for rapid growth and development (Fenner, 1987; Hanley *et al.*, 1996; Lemoine *et al.*, 2017).

To assess the relationship of seedling mortality with rainfall, maximum and minimum temperature, multiple regression analyses were conducted using data from NERIST site and found mixed results. While significant relationships between seedling mortality and rainfall ($p<0.005$) and seedling mortality and minimum temperature ($p<0.005$) were observed, non-significant relationships between seedling mortality and maximum temperature were depicted in the analysis (Fig. 5.15). This clearly indicates that *G. assamicus* seedlings are vulnerable to local climatic conditions such as precipitation and air temperature.

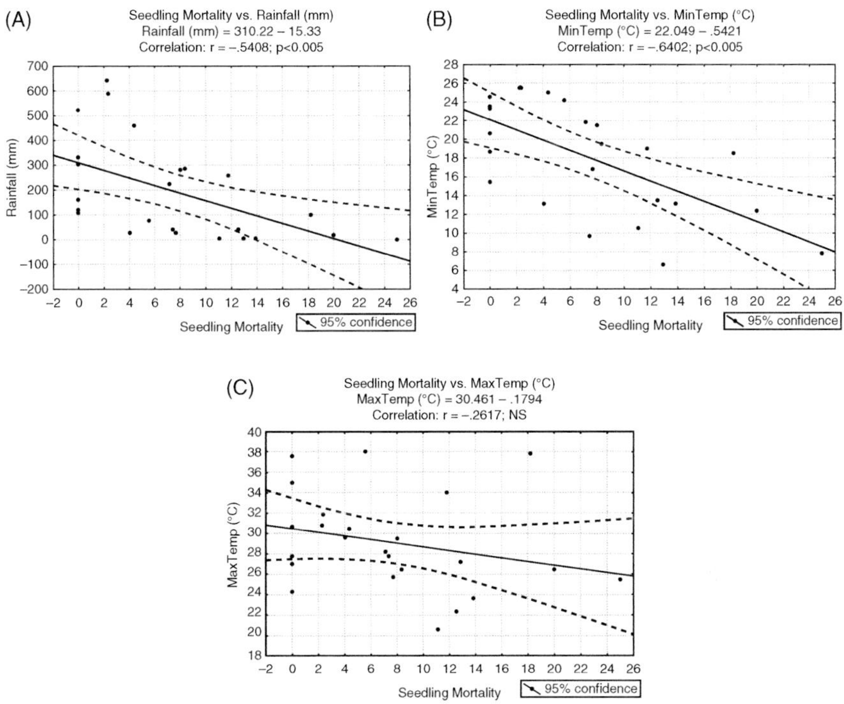

Fig. 5.15. Multiple regression analysis between seedling mortality and different environmental parameters in NERIST study site. (A) Regression between seedling mortality and rainfall; (B) Regression between seedling mortality and minimum temperature; and (C) Regression between seedling mortality and maximum temperature (NS = not significant).

5.4 Implications for Silvicultural and Conservation Practices

Plantation initiatives have been taken for many economically and ecologically important tree species either to meet demand or to conserve species or both. Nursery routines for these planting stocks for artificial regeneration are very important. Ecological studies of such species provide guidance for refining management interventions to improve establishment and growth. Several studies have been conducted throughout the world and have been reviewed by Baur (1968). In the present study, overall seedling performance was found to be superior in Dirang than the other sites. This may be due to a favourable microclimate throughout the year at Dirang site. Soil moisture may have played a significant role as regular watering was done in the nursery throughout the year. Seedling growth in terms of height and collar diameter was almost uniform among the seedlings of Moishing site and the transplanted seedlings at Dirang site. Similar geographic and climatic conditions prevailing at Dirang might have influenced such growth.

Once seedlings are transferred from nursery to the fields, attacks from many pests can no longer be prevented and are a major determinant of a species used and where they are planted (Hall, 1996). Our present study revealed that transplanted seedlings performed the best in Dirang area and least in NERIST botanical garden. This may be due to favourable growing conditions at Dirang as well as specific attributes of the seedlings which influence the physiological ability to utilize the environmental resources efficiently. On the other hand, poor growth performances in NERIST may be due to high irradiance levels at which photosynthetic systems is damaged causing bleaching of leaves (Oberbauer, 1985) or due to water limitation and high soil temperature. Whereas the altitudinal range of the native *G. assamicus* population is between 1,500–2,100 m asl and mean monthly temperature is remarkably lower, differential growth performances of transplanted seedlings in NERIST botanical garden may be attributed to lower altitude and higher mean monthly temperature. Similar results have also been reported in the case of *Shorea gardneri* in different ecological conditions (Ashton, 1990).

It was found that *G. assamicus* seedlings expressed maximum growth and development under partial canopy cover and in soil with higher moisture content in situ. Survival of transplanted seedlings was also found to be better under medium light irradiance levels at Dirang site. On the basis of this study, it may be suggested that a plantation of *G. assamicus* should be raised along river or stream bank where soil moisture is sufficient during maximum period of the year. Seedlings may be planted under medium irradiance levels along established forest or plantations with small to medium gaps.

5.5 References

Alexander, I., Ahmad, N. and See, L.S., 1992. The role of mycorrhizas in the regeneration of some Malaysian forest trees. *Philosophical Transactions of the Royal Society of London. Series B: Biological Sciences* 335(1275):379–388.

Andrus, R.A., Harvey, B.J., Rodman, K.C., Hart, S.J. and Veblen, T.T., 2018. Moisture availability limits subalpine tree establishment. *Ecology* 99(3): 567–575.

Antúnez, I., Retamosa, E.C. and Villar, R., 2001. Relative growth rate in phylogenetically related deciduous and evergreen woody species. *Oecologia* 128(2): 172–180.

Arianoutsou, M. and Thanos, C.A., 1996. Legumes in the fire-prone Mediterranean regions: an example from Greece. *International Journal of Wildland Fire* 6(2): 77–82.

Ashton, M.S., 1998. Seedling ecology of mixed-dipterocarp forest. In S. Appanah and J.M. Turnbull (eds) *A review of Dipterocarps: Taxonomy, Ecology and Silviculture.* Center for International Forestry Research, Bogor, pp. 89–98.

Ashton, P.M.S., 1990. Seedling response of *Shorea* species across moisture and light regimes in a Sri Lankan rain forest. PhD dissertation, Yale University, CT, USA.

Augspurger, C.K., 1983. Offspring recruitment around tropical trees: changes in cohort distance with time. *Oikos* 40(2): 189–196.

Augspurger, C.K. and Kelly, C.K., 1984. Pathogen mortality of tropical tree seedlings: experimental studies of the effects of dispersal distance, seedling density, and light conditions. *Oecologia* 61(2): 211–217.

Basset, Y., Charles, E. and Novotny, V., 1999. Insect herbivores on parent trees and conspecific seedlings in a Guyana rain forest. *Selbyana*: 146–158.

Baur, G.N., 1968. *The Ecological Basis of Rainforest Management.* Forestry Commision of New South Wales, Sydney.

Bertacchi, M.I.F., Amazonas, N.T., Brancalion, P.H., Brondani, G.E., de Oliveira, A.C., de Pascoa, M.A. and Rodrigues, R.R., 2016. Establishment of tree seedlings in the understory of restoration plantations: natural regeneration and enrichment plantings. *Restoration Ecology* 24(1): 100–108.

Bigras, F.J. and Dumais, D., 2005. Root-freezing damage in the containerized nursery: impact on plantation sites–A review. *New Forests* 30(2–3): 167–184.

Birhane, E., Kuyper, T.W., Sterck, F.J., Gebrehiwot, K. and Bongers, F., 2015. Arbuscular mycorrhiza and water and nutrient supply differently impact seedling performance of dry woodland species with different acquisition strategies. *Plant Ecology & Diversity* 8(3): 387–399.

Blennow, K. and Lindkvist, L., 2000. Models of low temperature and high irradiance and their application to explaining the risk of seedling mortality. *Forest Ecology and Management* 135(1–3): 289–301.

Bolte, A., Czajkowski, T., Cocozza, C., Tognetti, R., De Miguel, M., Pšidová, E., Ditmarová, Ĺ., Dinca, L., Delzon, S., Cochard, H. and Ræbild, A., 2016. Desiccation and mortality dynamics in seedlings of different European beech (*Fagus sylvatica* L.) populations under extreme drought conditions. *Frontiers in plant science* 7: 751.

Brokaw N.V.L., 1984. Gap phase regeneration in a Panamanian forest. In A.C. Chadwick and S.L. Sutton (eds) *Tropical Rainforest.* Leeds Philosophical and Literary Society, Central Museum, Calverley Street, Leeds, UK.

Brokaw, N.V.L., 1985. Tree falls, regrowth and community structure in tropical forests. In S.T.A. Pickekk and P.S. White (eds) *The Ecology of Natural Disturbance and Patch Dynamics.* Academic Press, Orlando, FL, pp. 53–69.

Brown, N.D. and Whitmore, T.C., 1992. Do dipterocarp seedlings really partition tropical rain forest gaps? *Philosophical Transactions of the Royal Society of London. Series B: Biological Sciences* 335(1275): 369–378.

Burslem D.F.R.P. 1996. Differential responses to nutrients, shade and drought among tree seedlings of lowland tropical forest in Singapore. In M.D. Swaine (ed.) *The Ecology of Tropical Forest Tree Seedlings.* MAB UNESCO Series Vol. 17, Parthenon, Paris, pp. 211–244.

Burslem, D.F.R.P., Grubb, P.J. and Turner, I.M., 1995. Responses to nutrient addition among shade-tolerant tree seedlings of lowland tropical rain forest in Singapore. *Journal of Ecology* 83(1): 113–122.

Canham, C.D., Berkowitz, A.R., Kelly, V.R., Lovett, G.M., Ollinger, S.V. and Schnurr, J., 1996. Biomass allocation and multiple resource limitation in tree seedlings. *Canadian Journal of Forest Research*, 26(9): 1521–1530.

Champion, H.G. and Seth, S.K., 1968. *A Revised Survey of the Forest Types of India.* Government of India Publication, Delhi.

Chang, C.Y., Unda, F., Zubilewich, A., Mansfield, S.D. and Ensminger, I., 2015. Sensitivity of cold acclimation to elevated autumn temperature in field-grown *Pinus strobus* seedlings. *Frontiers in Plant Science* 6: 165.

Chazdon, R.L., 1988. Sunflecks and their importance to forest understorey plants. *Advances in Ecological Research* 18: 1–63.

Chazdon, R.L. and Fetcher, N., 1984. Photosynthetic light environments in a lowland tropical rain forest in Costa Rica. *Journal of Ecology*: 553–564.

Choudhury, B.I., 2008. Regeneration ecology of *Gymnocladus assamicus* Kanjilal ex PC Kanjilal Leguminosae an economically important critically endangered tree species in Arunachal Pradesh. PhD thesis. Assam University, Silchar (India).

Choudhury, B.I., Khan, M.L. and Das, A.K., 2009. Seed dormancy and germination in *Gymnocladus assamicus*: An endemic legume tree from Northeast India. *Seed Science and Technology* 37(3): 582–588.

Clark, D.A. and Clark, D.B., 1984. Spacing dynamics of a tropical rain forest tree: evaluation of the Janzen-Connell model. *The American Naturalist* 124(6):769–788.

Clark, D.B. and Clark, D.A., 1989. The role of physical damage in the seedling mortality regime of a neotropical rain forest. *Oikos*: 55(2): 225–230.

Clark, D.B. and Clark, D.A., 1991. The impact of physical damage on canopy tree regeneration in tropical rain forest. *Journal of Ecology* 79(2): 447–457.

Connell, J.H., 1971. On the role of natural enemies in preventing competitive exclusion in some marine animals and in rain forest trees. *Dynamics of populations* 298: 312.

Coomes, D.A. and Grubb, P.J., 2000. Impacts of root competition in forests and woodlands: a theoretical framework and review of experiments. *Ecological monographs* 70(2): 171–207.

Dalling, J.W., Muller-Landau, H.C., Wright, S.J. and Hubbell, S.P., 2002. Role of dispersal in the recruitment limitation of neotropical pioneer species. *Journal of Ecology* 90(4): 714–727.

Dear, B.S. and Virgona, J.M., 1996. Legumes in low-input perennial pastures of southern Australia: historical role and future development. *New Zealand Journal of Agricultural Research* 39(4): 579–589.

De Candolle, A.P., 1825. Leguminosae. *Prodromus systematis naturalis regni vegetabilis* 2: 93–524.

De Steven, D., 1991. Experiments on mechanisms of tree establishment in old-field succession: seedling survival and growth. *Ecology* 72(3): 1076–1088.

Devi, K.A., 2004. Studies on plant diversity and regeneration of a few tree species in the Sacred Groves of Manipur. Ph.D thesis, North-Eastern Hill University, Shillong, India.

de Vogel E.F., 1980. *Seedlings of Dicotyledons: Structure, Development, Types: Description of 150 Woody Malesian Taxa*. Centre for Agricultural Publishing and documentation (PUDOC), Wageningen, Netherlands.

Duke J.A., 1970. Post irradiation woody seedlings. In H.T. Odum (ed.) *A Tropical Rain Forest*. US Atomic Energy Commission, Oak Ridge, pp. 189–192.

Fenner, M., 1987. Seedlings. *New Phytologist* 106: 35–47.

Flores, S., 1992. Growth and seasonality of seedlings and juveniles of primary species of a cloud forest in northern Venezuela. *Journal of Tropical Ecology* 8(3): 299–305.

Foster, S.A., 1986. On the adaptive value of large seeds for tropical moist forest trees: a review and synthesis. *The Botanical Review* 52(3): 260–299.

Fredericksen, T.S. and Licona, J.C., 2000. Invasion of non-commercial tree species after selection logging in a Bolivian tropical forest. *Journal of Sustainable Forestry* 11(3): 113–123.

Frehner, M., 2002. Untersuchungen über den Einfluss unterschiedlicher Kleinstandorte und der Pflanztechnik auf Fichtenpflanzungen in subalpinen Lawinenschutzwäldern (Doctoral dissertation, ETH Zurich).

Gardiner, E.S. and Hodges, J.D., 1998. Growth and biomass distribution of cherrybark oak (*Quercus pagoda* Raf.) seedlings as influenced by light availability. *Forest Ecology and Management* 108(1–2): 127–134.

Garwood, N.C., 1983. Seed germination in a seasonal tropical forest in Panama: a community study. *Ecological monographs* 53(2): 159–181.

Garwood, N.C., 1996. Functional morphology of tropical tree seedlings. In M.D. Swaine (ed.) *The Ecology of Tropical Forest Tree Seedlings*. Man and Biosphere Series. UNESCO/Parthenon, Paris/Carnforth, 17, pp. 59–118.
Gómez-Aparicio, L., Pérez-Ramos, I.M., Mendoza, I., Matias, L., Quero, J.L., Castro, J., Zamora, R. and Maranon, T., 2008. Oak seedling survival and growth along resource gradients in Mediterranean forests: implications for regeneration in current and future environmental scenarios. *Oikos* 117(11): 1683–1699.
Grau, H.R., 2002. Scale-dependent relationships between treefalls and species richness in a neotropical montane forest. *Ecology*, 83(9): 2591–2601.
Hall, J.B., 1996. Seedling ecology and tropical forestry. In M.D. Swaine (ed.) *The Ecology of Tropical Forest Tree Seedlings*. Man and Biosphere Series. UNESCO/Parthenon, Paris/Carnforth, 17, pp. 139–159.
Halvorson, J.J., Black, R.A., Smith, J.L. and Franz, E.H., 1991. Nitrogenase activity, growth and carbon and nitrogen allocation in wintergreen and deciduous lupin seedlings. *Functional Ecology* 5. pp. 554–561.
Hanley, M.E., Fenner, M. and Edwards, P.J., 1996. The effect of mollusc grazing on seedling recruitment in artificially created grassland gaps. *Oecologia* 106(2): 240–246.
Hladik, A. and Miquel, S., 1990. Seedling types and plant establishment in an African rain forest. In K.S. Bawa and M. Hadley (eds) *Reproductive Ecology of Tropical Forest Plants*. UNESCO/Parthenon, Paris/Carnforth, pp. 261–282.
Hogberg, P., 1986. Soil nutrient availability, root symbioses and tree species composition in tropical Africa: a review. *Journal of Tropical Ecology* 2(4): 359–372.
Hunt, W.F. and Halligan, G., 1981. Growth and developmental responses of perennial ryegrass grown at constant temperature. I. Influence of light and temperature on growth and net assimilation. *Functional Plant Biology* 8(2): 181–190.
Hunziker, U. and Brang, P., 2005. Microsite patterns of conifer seedling establishment and growth in a mixed stand in the southern Alps. *Forest Ecology and Management* 210(1–3): 67–79.
Janzen, D.H., 1970. Herbivores and the number of tree species in tropical rain forests. *American Naturalist* 104: 501–528.
Johnson, D.J., Condit, R., Hubbell, S.P. and Comita, L.S., 2017. Abiotic niche partitioning and negative density dependence drive tree seedling survival in a tropical forest. *Proceedings of the Royal Society B: Biological Sciences* 284(1869): 2017–2210.
Khan, M.L., 2004. Effects of seed mass on seedling success in *Artocarpus heterophyllus* L., a tropical tree species of north-east India. *Acta Oecologica* 25(1–2): 103–110.
Khan, M.L. and Shankar, U., 2001. Effect of seed weight, light regime and substratum microsite on germination and seedling growth of *Quercus semiserrata* Roxb. *Tropical Ecology* 42(1): 117–125.
Khan, M.L., Rai, J.P.N. and Tripathi, R.S., 1986. Regeneration and survival of tree seedlings and sprouts in tropical deciduous and sub-tropical forests of Meghalaya, India. *Forest Ecology and Management* 14(4): 293–304.
Kim, T.N. and Underwood, N., 2015. Plant neighborhood effects on herbivory: damage is both density and frequency dependent. *Ecology* 96(5): 1431–1437.
Kitajima, K. and Fenner, M., 2000. Ecology of seedling regeneration. In M. Fenner (ed) *Seeds: The Ecology of Regeneration in Plant Communities* (2nd edn). CAB International, Wallingford, UK, pp. 331–359.
Kohyama, T. and Hara, T., 1989. Frequency distribution of tree growth rate in natural forest stands. *Annals of Botany* 64(1): 47–57.
Kukkonen, M., Rita, H., Hohnwald, S. and Nygren, A., 2008. Treefall gaps of certified, conventionally managed and natural forests as regeneration sites for Neotropical timber trees in northern Honduras. *Forest Ecology and Management* 255(7): 2163–2176.

Kumar R., Singh, A.K. and Abbas, S.G., 1994. Change in population structure of some dominant tree species of dry Peninsular Sal Forest. *Indian Forester* 120(4): 343–348.

Lambers, H.A.N.S. and Poorter, H., 1992. Inherent variation in growth rate between higher plants: a search for physiological causes and ecological consequences. *Advances in ecological research* 23: 187–261.

Latorre, L., Larrinaga, A.R. and Santamaría, L., 2013. Combined impact of multiple exotic herbivores on different life stages of an endangered plant endemism, *Medicago citrina*. *Journal of Ecology* 101(1): 107–117.

Laughlin, D.C. and Clarkson, B.D., 2018. Tree seedling survival depends on canopy age, cover and initial composition: trade-offs in forest restoration enrichment planting. *Ecological Restoration* 36(1): 52–61.

Lee, D.W., Baskaran, K., Mansor, M., Mohamad, H. and Yap, S.K., 1996. Irradiance and spectral quality affect Asian tropical rain forest tree seedling development. *Ecology* 77(2): 568–580.

Lemoine, N.P., Doublet, D., Salminen, J.P., Burkepile, D.E. and Parker, J.D., 2017. Responses of plant phenology, growth, defense, and reproduction to interactive effects of warming and insect herbivory. *Ecology* 98(7): 1817–1828.

Li, Y., Ran, W., Zhang, R., Sun, S. and Xu, G., 2009. Facilitated legume nodulation, phosphate uptake and nitrogen transfer by arbuscular inoculation in an upland rice and mung bean intercropping system. *Plant and Soil* 315(1–2): 285–296.

Lieberman, D. and Li, M., 1992. Seedling recruitment patterns in a tropical dry forest in Ghana. *Journal of Vegetation Science* 3(3): 375–382.

Lusk, C.H. and Laughlin, D.C., 2017. Regeneration patterns, environmental filtering and tree species coexistence in a temperate forest. *New Phytologist* 213(2): 657–668.

Maron, J.L. and Crone, E., 2006. Herbivory: effects on plant abundance, distribution and population growth. *Proceedings of the Royal Society B: Biological Sciences* 273(1601): 2575–2584.

McLeod, K.W. and Murphy, P.G., 1977. Establishment of Ptelea trifoliata on Lake Michigan sand dunes. *American Midland Naturalist* 97(2): 350–362.

Menezes, T.G.C., Rodal, M.J.N., Lima, A.L.A.D., Lima, L.R., Pinto, M.A.D.S. and Melo, A.L.D., 2018. Do seedling functional groups reflect ecological strategies of woody plant species in Caatinga? *Acta Botanica Brasilica* 32(1): 20–27.

Metz, M.R., Sousa, W.P. and Valencia, R., 2010. Widespread density-dependent seedling mortality promotes species coexistence in a highly diverse Amazonian rain forest. *Ecology* 91(12): 3675–3685.

Michelsen, A., 1992. Mycorrhiza and root nodulation in tree seedlings from five nurseries in Ethiopia and Somalia. *Forest Ecology and Management* 48(34): 335–344.

Moles, A.T. and Westoby, M., 2004. Seedling survival and seed size: a synthesis of the literature. *Journal of Ecology* 92(3): 372–383.

Mori, T., Nakashizuka, T., Sumizono, T. and Yap, S.K., 1990. Growth and photosynthetic responses to temperature in several Malaysian tree species. *Journal of Tropical Forest Science* 3(1): 44–57.

Mueller-Dombois D., Jacobi J.D., Cooray R.G. and Balakrishnan N., 1980. *Ohia Rainforest Study: Ecological Investigations of the Ohia Die-Back Problem in Hawaii*. Miscellaneous Publications, 183. Hawaii Institute of Tropical agriculture and Human resources, Honolulu, Hawaii.

Mueller-Dombois, D., Canfield, J.E., Holt, R.A. and Buelow, G.P., 1983. Tree-group death in North American and Hawaiian forests: a pathological problem or a new problem for vegetation ecology? *Phytocoenologia* 11(1): 117–137.

Myers, J.H. and Sarfraz, R.M., 2017. Impacts of insect herbivores on plant populations. *Annual review of entomology*, 62: 207–230.

Myster, R.W., 2002. Foliar pathogen and insect herbivore effects on two landslide tree species in Puerto Rico. *Forest Ecology and Management*, 169(3): 231–242.

Ng, F.S.P., 1978. Strategies of establishment in Malayan forest trees. In P.B. Tomlinson and M.H. Zimmerman (eds) *Tropical Trees as Living Systems*. Cambridge University Press, Cambridge, pp. 129–162.

Nilsen, E.T., Rundel, P.W. and Sharifi, M.R., 1981. Summer water relations of the desert phreatophyte *Prosopis glandulosa* in the Sonoran desert of southern California. *Oecologia* 50(2): 271–276.

Oberbauer, S.F., 1985. Plant water relation of selected species in wet and dry tropical lowland forest in Costa Rica. *Revista de Biologi Tropical* 33: 137–142.

Osunkjoya, O.O., Ash, J.E., Hopkins, M.S. and Graham, A.W., 1992. Factors affecting survival of tree seedlings in North Queensland rainforests. *Oecologia* 91(4): 569–578.

Packer, A. and Clay, K., 2000. Soil pathogens and spatial patterns of seedling mortality in a temperate tree. *Nature* 404(6775): 278.

Pearcy, R.W., 1987. Photosynthetic gas exchange responses of Australian tropical forest trees in canopy, gap and understory micro-environments. *Functional Ecology* 1(3): 169–178.

Pearcy, R.W., 1990. Sunflecks and photosynthesis in plant canopies. *Annual Review of Plant Biology* 41(1): 421–453.

Pernot, C., Thiffault, N. and DesRochers, A., 2019. Root system origin and structure influence planting shock of black spruce seedlings in boreal microsites. *Forest Ecology and Management* 433: 594–605.

Piper, C.S., 1944. *Soil and Plant Analysis*. John Wiley & Sons, New York.

Pompa, J. and Bongers, F., 1988. The effect of canopy gaps on growth and morphology of seedlings of rain forest species. *Oecologia* 75(4): 625–632.

Purdie, R.W., 1977. Early stages of regeneration after burning in dry sclerophyll vegetation. II. Regeneration by seed germination. *Australian Journal of Botany* 25(1): 35–46.

Quero, J.L., Villar, R., Marañón, T. and Zamora, R., 2006. Interactions of drought and shade effects on seedlings of four *Quercus* species: physiological and structural leaf responses. *New Phytologist* 170(4): 819–834.

Rao, P.B. and Singh, S.P., 1985. Response breadths on environmental gradients of germination and seedling growth in two dominant forest tree species of central Himalaya. *Annals of Botany* 56(6): 783–794.

Reeves, J.T., Ezell, A.W., Hodges, J.D., Schultz, E.B. and Self, A.B., 2016. Influence of flooding, freezing, and American beaver herbivory on survival of planted oak seedlings. *Proceedings of the 18th biennial southern silvicultural research conference*. e-Gen. Tech. Rep. SRS-212. Asheville, NC: US Department of Agriculture, Forest Service, Southern Research Station 614(212): 465–468.

Rodrigues, R.S., 2015. Are seedlings diagnostic in *Neotropical Entada* (Leguminosae)? Seedling morphology supports the reinstatement of *Entada polyphylla*. *Phytotaxa* 220(3): 287–294.

Roff, D. (ed.), 1992. *Evolution of Life Histories: Theory and Analysis*. Chapman & Hall, New York.

Rundel, P.W., 1989. Ecological success in relation to plant form and function in the woody legumes. In C.H. Stirton and J.L. Zarucchi (eds) *Advances in Legume Biology*. Monographs in Systematic Botany, Missouri Botanical Garden, 29: 377–398.

Runkle, J.R. and Yetter, T.C., 1987. Treefalls revisited: gap dynamics in the southern Appalachians. *Ecology* 68(2): 417–424.

Sakai, A. and Larcher, W., 2012. *Frost Survival of Plants: Responses and Adaptation to Freezing Stress*, vol. 62. Springer Science & Business Media, New York.

Sankey, T.T., Montagne, C., Graumlich, L., Lawrence, R. and Nielsen, J., 2006. Lower forest–grassland ecotones and 20th century livestock herbivory effects in northern Mongolia. *Forest Ecology and Management* 233(1): 36–44.

Sasaki, S. and Mori, T., 1981. Growth responses of dipterocarp seedlings to light. *Malaysian Forester* 44(2/3): 319–345.

Scariot, A., 2000. Seedling mortality by litterfall in Amazonian forest fragments 1. *Biotropica* 32(4a): 662–669.

Smits, W.T.M., 1983. Dipterocarps and mycorrhiza. An ecological adaptation and a factor in forest regeneration. *Flora Malesiana Bulletin* 36(1): 3926–3937.

Snook, L.C., 1982. Tagasaste (tree lucerne) *Chamaecytisus palmensis*: a shrub with high potential as a productive fodder crop. *Journal of the Australian Institute of Agricultural Science* 48: 209–213.

Steege H., 1990. *A Monograph of Wallaba, Mora and Greenheart.* (Tropenbos Technical Series 5). The Tropenbos Foundation, Wageningen, The Netherlands.

Stewart, G.R., Gracia, C.A., Hegarty, E.E. and Specht, R.L., 1990. Nitrate reductase activity and chlorophyll content in sun leaves of subtropical Australian closed-forest (rainforest) and open-forest communities. *Oecologia* 82(4): 544–551.

Swaine, M.D. and Whitmore, T.C., 1988. On the definition of ecological species groups in tropical rain forests. *Vegetatio*, 75(1–2): 81–86.

Swaine, M.D., Lieberman, D. and Hall, J.B., 1990. Structure and dynamics of a tropical dry forest in Ghana. *Vegetatio* 88(1): 31–51.

Swinfield, T., Lewis, O.T., Bagchi, R. and Freckleton, R.P., 2012. Consequences of changing rainfall for fungal pathogen-induced mortality in tropical tree seedlings. *Ecology and Evolution* 2(7): 1408–1413.

Tanouchi, H., Sato, T. and Takeshita, K., 1994. Comparative studies on acorn and seedling dynamics of four *Quercus* species in an evergreen broad-leaved forest. *Journal of Plant Research* 107(2): 153–159.

Ter Heerdt, G.N., Veen, C.G., Van der Putten, W.H. and Bakker, J.P., 2017. Effects of temperature, moisture and soil type on seedling emergence and mortality of riparian plant species. *Aquatic Botany* 136: 82–94.

Turnbull, M.H., 1991. The effect of light quantity and quality during development on the photosynthetic characteristics of six Australian rainforest tree species. *Oecologia* 87(1): 110–117.

Turner, I.M., 1990. The seedling survivorship and growth of three *Shorea* species in a Malaysian tropical rain forest. *Journal of Tropical Ecology* 6(4): 469–478.

Watkinson, A.R., 1996. Plant population dynamics. In: M.J. Crawley (ed.), *Plant Ecology*: pp. 359–400. Blackwell Scientific Publications, Oxford, UK.

Whitmore, T. C., 1975. *Tropical Rain Forest of Far East.* Claredon Press, Oxford, p. 282.

Whitmore, T.C., 1984. *Tropical Rain Forests of the Far East* (2nd edn). Clarendon Press, Oxford.

Whitmore T.C., 1990. *An Introduction to Tropical Rain Forests.* Clarendon Press, Oxford.

Whitmore, T.C., 1996. A review of some aspects of tropical rain forest seedling ecology with suggestions for further enquiry. *Man and the Biosphere Series* 17: 3–40.

Williams, P.R., Congdon, R.A., Grice, A.C. and Clarke, P.J., 2005. Effect of season of burning and removal of herbaceous cover on seedling emergence in a eucalypt savanna of north-eastern Australia. *Austral Ecology* 30(5): 491–496.

Yahara, T., Javadi, F., Onoda, Y., de Queiroz, L.P., Faith, D.P., Prado, D.E., Akasaka, M., Kadoya, T., Ishihama, F., Davies, S. and Slik, J.F., 2013. Global legume diversity assessment: concepts, key indicators, and strategies. *Taxon* 62(2): 249–266.

6

Conservation and Management of Plant Biodiversity in the Eastern Himalayan Region with Emphasis on Soap Pod Tree (*Gymnocladus assamicus*)

6.1 Introduction

Biodiversity or biological diversity is defined as 'the variability among living organisms from all sources, including interalia, terrestrial, marine and other aquatic ecosystems and the ecological complexes of which they are part; this includes diversity within species, between species and of ecosystems' (definition adopted following the Convention on Biological Diversity – CBD – at the Earth Summit, Rio de Janeiro, 1993). Biodiversity is a crucial component of planet earth and a basis of human existence. The values of biodiversity are tremendous and provide enormous services to our livelihood. Biodiversity sustains several resources upon which communities, nations and future generations depend. It also provides services to our health, agriculture and businesses. Therefore, loss of biodiversity in any form has serious economic and social costs for any nation.

However, in recent years we have lost biodiversity in all ecosystems at an unprecedented rate. Global biodiversity is under threat and facing tremendous pressure due to various natural and anthropogenic activities. Global warming is increasingly threatening to different species which may face extinction pressure in the near future. As a result, several species of organisms (microbes, plants and animals) are depleting at a rapid pace in an unprecedented way. Biodiversity loss has been identified as a global phenomenon and is accelerating in scale and scope alarmingly (Wood, 2000). While the extinction of species in nature is spontaneous, the rate of present-day extinction is faster than at any point in the past. The rate of extinction has significantly increased in recent decades and is estimated to be 1,000–10,000 times higher than the natural extinction rate (Chivian and

Bernstein, 2008). Several agencies are closely monitoring the status of biodiversity using different approaches to develop strategies for effective conservation. The International Union for Conservation of Nature (IUCN) assessed 71,576 mostly terrestrial and freshwater species (until March 2014) and found 860 extinct or extinct in the wild, 21,286 threatened and 4,286 deemed critically endangered (IUCN, 2019a).

Extinction of species is considered as one of the biggest threats to humankind. Notably, the threat to most of the species concerned is primarily due to human activities such as habitat degradation, fragmentation, resource exploitation and global climate change. With the current extinction rate, 30–50% of all species will be lost by the midcentury (Thomas *et al.*, 2004) if effective conservation and management strategies of biological diversity are not adopted. Therefore, effective conservation and management are urgently needed to face the challenges of climate change, sustainable development and food security.

6.2 Biodiversity in India

India is Asia's second largest nation and one of the most densely populated countries in the world. The total geographical area of India is around 3,287,263 sq. km with a land frontier of around 15,200 km and a coastline of around 7,517 km. The country is home to a rich wealth of biological diversity in its forests, wetlands and marine ecosystems. India possesses a distinct identity, not only because of its geography, history and culture but also because of the great diversity of its natural ecosystems and large climatic and topographic gradient. India is a megadiverse country with only 2.4% of the world's land area, and accounts for 7–8% of all recorded species, including over 45,000 species of plants and 91,000 species of animals (IUCN, 2019b). The Indian subcontinent has diverse forest ecosystems which have been classified into six 'major groups' ranging from tropical to alpine, and 16 'groups' and more than 200 'group categories' (Singh and Chaturvedi, 2017). Champion and Seth (1968) classified present-day India into 16 major forest types which are again subdivided into 221 minor types. This classification provides valuable information, especially from an administrative perspective, and is largely used in several developmental, policy-making and conservation-related issues.

Based on principal criteria such as species endemism and marine ecosystems within its borders, India is one of the 17 megadiversity countries in the world (UNEP-WCMC, 2014). The country has many endemic plant species which represent about 11% of the world's flora in about 2.4% of the global land mass. More specifically, about 28% of the total flora and 33% of angiosperms found in India are endemic (Chitale *et al.*, 2014) (Table 6.1). India also has 4 of the 35 globally identified biodiversity hotspots – namely, Himalaya, Indo-Burma, Western Ghats and Sri Lanka (Mittermeier *et al.*, 2011; Williams *et al.*, 2011). A biodiversity hotspot is determined based on the high percentage of endemic vascular plants (minimum number 1,500) which is also threatened (must have 30% or less of its

Table 6.1. Plant diversity in India at a glance (Wildlife Institute of India, 2007).

Group	No. of species
Angiosperms (flowering plants)	17,000
Gymnosperms	64
Pteridophytes	1,022
Ferns	1,200
Bryophytes	2,584
Lichens	1,600
Fungi	2,300
Algae	2,500
Bacteria	850
Orchids	1,100
Endemic plants	6,850
Plants of ethnobotanical interest	800

original natural vegetation) (Conservation International, 2019). These hotspots are the storehouse of several rare, endangered, endemic and primitive species of plants and animals and other living organisms. Northeast India is the confluence of three different geological origins and harbours a very rich biota which has an association from three different biogeographic realms (Myers, 1988). The region is one of the highest biodiversity-rich areas in India. All the northeastern states (Arunachal Pradesh, Assam, Manipur, Meghalaya, Mizoram, Nagaland, Sikkim and Tripura) make up just 7.98% (2,62,179 sq. km) of the total geographical area of the country, but have about 25% of the country's total forest area, supporting about 30% of the total growing stock of the forest. Nearly 64% of the total geographical area has forest cover (State of Forest Report, 2017). The region has more than 65% forest cover of the total geographical area compared to just above 21% nationally. The state of Arunachal Pradesh has the highest land mass (83,743 sq. km) under forest cover which is about 80% of the total land area of the state (State of Forest Report, 2017). Mizoram has the highest forest cover (86%) while Assam has the least (about 36%) as per the latest assessment (Fig. 6.1) (State of Forest Report, 2017).

6.3 Biodiversity in Northeast India

The northeast Indian region can be broadly differentiated into the eastern Himalaya to the north, the northeast hills to the south and the Brahmaputra River basin (BRB) in between (Pawar *et al.*, 2007). Northeast India has a comparatively complex biogeography due to its age, unique plate tectonics and palaeoclimatic history, its location at the confluence of distinct realms having a wide physiognomic range (e.g. altitude ranging from ca. 100 to >7000 m above sea level)

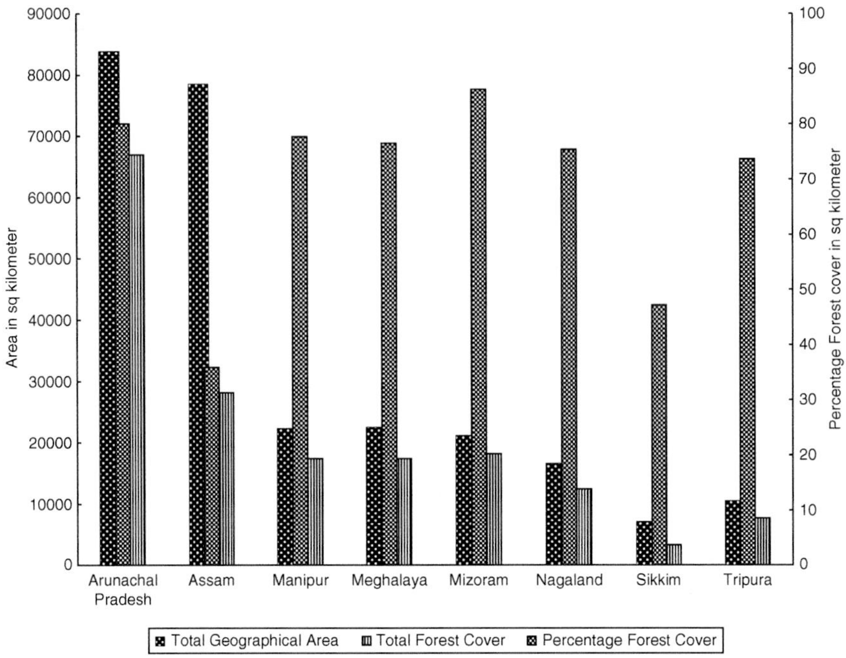

Fig. 6.1. Forest cover of the northeastern states as per 2017 assessment (data adopted from the State of Forest Report, 2017).

and habitat diversity (Olson *et al.*, 2001). Major forest ecosystems of northeast India based on elevation (above sea level) are: alpine temperate (above 3,500 m); subtropical pine (1,000–3,500 m); subtropical broad-leaved (900–1900 m); tropical wet evergreen (up to 900 m); and tropical semievergreen (up to 600 m) (Poffenberger *et al.*, 2006). Tropical forest vegetation is abundant in the region and typically occurs at elevations up to 900 m above sea level. The region also harbours evergreen and semievergreen rainforests, moist deciduous monsoon forests, riparian forests, swamps and grasslands. Evergreen rain forests occur in the plains of the Assam Valley, Barak Valley, the foothills of the eastern Himalayas and the lower parts of the Naga Hills (Meghalaya, Mizoram and Manipur) where rainfall exceeds 2,300 mm per annum. Lowland evergreen rainforest areas in the region are characterized by the presence of frequent buttressing, large woody climbers and dominance of two Dipterocarpaceae species, *Dipterocarpus macrocarpus* and *Shorea assamica* (Procter *et al.*, 1998).

Ethnocultural diversity of the region is enormous and it is home to approximately 225 tribes out of 450 in India (Chatterjee *et al.*, 2006). The culture and customs of these tribes have an important role in understanding biodiversity conservation and management issues (Chatterjee *et al.*, 2006). The lowland and montane moist to wet tropical evergreen forests of the region are considered to be the northernmost limit of true tropical rainforests in the world (Procter *et al.*, 1998).

Rhododendrons are an important group of plants belonging to the family Ericaceae occur throughout the Himalayan region and about 98% of Indian rhododendron species occur in the Himalayan region (Paul *et al.*, 2005). Very high flowering plant diversity is also a general characteristic of the northeastern states (Table 6.2). At least 46 *Rhododendron* species have been categorized as rare or threatened in the eastern Himalayan region of India (Paul *et al.*, 2005). Northeastern region of India still remains fully unexplored and new species are described frequently.

Northeastern region region is also very rich in crop genetic resources with several species of wild crop relatives of economic importance. Studies have demonstrated significant levels of genetic diversity among a few traditional crop varieties of the region such as taro (Lakhanpaul *et al.*, 2003), ginger (Sajeev *et al.*, 2011), rice (Choudhury *et al.*, 2013) and citrus (Barbhuiya *et al.*, 2016a). The region has been identified by the Indian Council of Agricultural Research (ICAR) as a centre of rice germplasm and the National Bureau of Plant Genetic Resources (NBPGR, India) as a region rich in wild relatives of crop plants in India (Chatterjee *et al.*, 2006). The region is well known for its unique flora and fauna, high endemism and rare, endangered and primitive species. The 'Himalaya hotspot' is home to around 10,000 species of plants of which about 32% are endemic (Conservation International, 2006).

Different primitive and fossil plants have also been documented in the region from different periods of time. Bhattacharyya (1967) described fossils of *Nipa* and *Calophyllum* from Lower to Middle Eocene beds near Cherrapunji and Latrayengew in Meghalaya. Lakhanpal (1954, 1955) reported leaf impressions of *Nelumbium* sp., *Trema garoensis*, *Neolitisia sahnii*, *Grewia foxii* and *Bombacites orientalis* from Middle Eocene beds near Damalgiri in Garo Hills, Meghalaya. Report of palm leaf *Sabalites* from Laitryngew near Shillong by Bose and Sah (1964) is also remarkable. Bhattacharyya (1983, 1985) described a number of angiospermic leaf, flower and fruit impressions from different parts of Meghalaya which are comparable with modern taxa (Table 6.3). Several Neogene flora (45 genera of 20 families) from northeast India (Assam-Arakan Basin) are also indicative of

Table 6.2. Floral diversity among the northeastern states of India (adapted from Chatterjee *et al*, 2006).

State	Species richness (flowering plants)
Arunachal Pradesh	±5,000
Assam	±3,010
Manipur	±2,500
Meghalaya	±3,500
Mizoram	±2,200
Nagaland	±2,250
Sikkim	±4,500
Tripura	±1,600

Table 6.3. Angiospermic leaf, flower and fruit impressions from different parts of Meghalaya (adapted from Singh *et al.*, 2000).

Location	Fossil plant	Similarity with modern taxa
West Daranggiri	*Nelumbo*	*Nelumbo nucifera*
Rongrenggiri and Nangalbibra	*Nipa sahnii*	*Nipa*
	Nelumbo nangalensis	*Nelumbo*
	Litsea sp	*Litsea*
	Phoebe sublanceolata	*Phoebe*
	Arocarpus garoensis	*Arocarpus*
	Triumfetta rhomboideacarpa	*Triumfetta*
	Osmanthus eocenicus	*Osmanthus*
	Ligustrum turaensis	*Ligustrum*
	Antholithes oleaceaeformis	Not known
	Antholithes campanulatum	Not known
	Poacites sp.	Not known
Garo Hills	*Leguminocarpon desmodioedes*	*Desmodium triquetrum*
	L. derrisoides	*Derris cuneifolia*
	L. mellettiodes	*Millettia rubiginosa*
	L. pongamioedes	*Pongamia pinnata*
	L. albizioides	*Albizia lucida*

warm tropical rainforest conditions during the Neogene (Singh *et al.*, 2000). The presence of *Nipa* in the Garo Hills of Meghalaya and Afzelia-Intsia in Cachar Hills of Assam also suggests the existence of coastal conditions in the area during the Neogene period (Singh *et al.*, 2000). Reports of megafossil leaves of *Ziziphus* and *Dioscorea* from the Siang district of Arunachal Pradesh come from Singh and Prakash (1980) and of small American elements such as *Adenocaulon*, *Oxybaphus*, *Podophyllum*, *Meconopsis* Sect. *Stylopodium*, *Liquidumber*, *Gnetum*, *Lardizabala*, *Monotropa uniflora*, *Brassenia* and *Mitreola* from Vishnu-Mittre (1969): these are from northeast India and are significant in this context. A number of primitive angiosperms such as *Altingia* sp., *Crylopsis himalayana*, *Distylium* sp., *Exbucklandia populnea*, *Holbolllia latifolia* var. *angustifolia*, *Houttuynia cordata*, *Magnolia pealiana*, *M. griffithii*, *M. gustavii*, *M. pterocarpa*, *Manglietia caveana*, *Myrica esculanta*, *Talauma hodgsonii* and *Tetracentron sinensis* are represented in the flora of northeast India and occurrence of these primitive species makes this area a cradle of flowering plants (Takhtajan, 1969).

Several initiatives have been put in place to protect the rich biodiversity of the region, including protected areas. A protected area is especially dedicated to the protection and maintenance of biological diversity and of natural and associated cultural resources, managed through legal or other effective means. The richness of biological diversity in northeast India has been given priority in declaring protected areas of different criteria for effective conservation and to protect different habitats and several rare/endangered species, and provide a safe place for other native populations. Several protected areas consisting of biosphere reserves, national parks and wildlife sanctuaries lie within the boundaries of northeast India.

6.4 State-wise Biodiversity Status in Northeast India

It is important to systematically evaluate the biodiversity status of different administrative units (states in the current perspective) to develop effective conservation strategies and future utilization.

6.4.1 Arunachal Pradesh

Arunachal Pradesh is the extreme northeastern state of India with a land area of 8.37 million ha and is the largest state among the other northeastern states. It consists of 25 administrative districts (Government of Arunachal Pradesh, 2019) and harbours amazingly high biodiversity. The species richness of the state is very high with 4,000–5,000 species of vascular plants per 10,000 sq. km and it is thus considered the second richest region of biodiversity in global context (Valdiya, 2002). Diverse and wonderful habitats within the state provide amazing vegetation that nourishes more than 5,000 species of flowering plants, 600 species of orchids, 89 species of bamboos, 18 species of canes, 400 species of ferns, 24 species of gymnosperms and innumerable species of algae, fungi, lichens, bryophytes and microorganisms (Hegde, 2002). Faunal diversity is also very high with more than 214 species of mammals, 770 species of birds, 83 species of reptiles, 130 species of fish and 7 species of non-human primates, along with numerous species of insects (Hegde, 2002). The cultural diversity of the state is incredible. It is home to 26 major Indo-Mongoloid tribes and 125 subtribes and clans who have diverse social structures and idiosyncratic acculturations (Bhagawati, 2002). Arunachal Pradesh has a forest area of 66,964 sq. km (constituting 79.96% of the total geographical area) of which 20,721 sq. km is very dense forest and 30,995 sq. km is moderately dense forest (State of Forest Report, 2017). The Monpa tribe (Fig. 6.2) dominates the West Kameng and Tawang districts of Arunachal Pradesh where our study was conducted. Tribal people in the state are intricately connected with the forest and often live in small villages within the forest (Fig. 6.3). Accordingly, people largely depend on forest resources such as bamboo, timber, firewood and minor forest produce (Fig. 6.4).

Diverse phytogeographical position along with irregular and undulating topography with wide variation in climate and soil have resulted in distinct vegetation and several forest types within the state of Arunachal Pradesh. Kaul and Haridasan (1987) classified the forest of the region into six major types within four climatic zones. The five major vegetation types are tropical, subtropical, temperate, subalpine and alpine. The state is home to several interesting species such as *Aeginetia indica*, *Arabidopsis himalaica*, *A. mollissima*, *Balanophora dioica*, *Boschiniakia himalaica*, *Raopalocnemis phalloides*, *Sapria himalayana*, as well as several species of

Fig. 6.2. A tribal man belonging to Monpa tribe in Dirang area of West Kameng district, Arunachal Pradesh (photo: Baharul I. Choudhury).

ferns including majestic tree ferns, orchids, primulas, etc. Arunachal Himalaya is home to around 85% of Indian rhododendrons and constitutes an important and interesting group of plants both ecologically and economically (Paul *et al.*, 2005). Hajra *et al.* (1996) listed several endemic, rare and endangered plant species from Arunachal Pradesh. A few endemic species from the state include *Acer oblongum* var. *microcarpum*, *Acer sikkimensis* var. *serrulatum*, *Aconogonum pangianum*, *Begonia aborensis*, *Begonia scintillans*, *Begonia tessaricarpa*, *Capparis pachyphylla*, *Coptis teeta*, *Hedychium longipedunculatum*, *Illicium cambodianum*, *Lysimachia congestiflora* var. *santapaui*, *Maesa arunachalensis*, *Merrilliopanax listeri*, *Paphiopedilum wardii*, *Pholidota watti*, *Primula subansirica*, *Pternopetalum* senii, *Pueraria bella*, *Rhododendron santapaui*, *Rhynchoglossum lazulinum*, *Schefflera venulosa*, *Tricarpelema glanduliferum* and *Wallichiana triandra*. Gajurel (2002), in an extensive study, described 29 *Piper* species with one new record for India and many lesser known species. He has also described three new *Piper* species from the state of Arunachal Pradesh. Arunachal Pradesh is also home to flagship animal

Fig. 6.3. A village within the forest area in Aalo region of West Siang district, Arunachal Pradesh (photo: Baharul I. Choudhury).

species such as mithun (*Bos frontalis*) (Fig. 6.5) and yak (*Bos grunniens*) which have tremendous economic and cultural significance in the state.

6.4.2 Assam

Assam is the second largest state in northeast India geographically after Arunachal Pradesh (78,438 sq. km) and the most densely populated state of the region. The state is located between 24°08' and 27°9' N latitudes and 89°42' and 96°10' E longitudes. As per the revised survey, Assam has seven major forest types viz. tropical wet evergreen forests, tropical semievergreen forests, tropical moist deciduous forests, subtropical broadleaf hill forests, subtropical pine forests, littoral and swamp forests, and grassland and savannahs (Champion and Seth, 1968).

Assam is also known for its important conservation sites such as Kaziranga National Park (famous for one horn rhino, *Rhinoceros unicornis*) and Manas Biosphere reserve for tigers. According to the available records and enumeration lists, 3,854 species of flowering plants have been recorded in the state (Chatterjee *et al.*, 2006; Anonymous, 2019). Although the exact number of plants is still uncertain, a high diversity of plant resources is expected from different eco-climatic zones of the state. The vegetation of the region is characterized predominantly by dense evergreen, semievergreen and moist deciduous forest in addition to degraded forest, bamboo and gregarious

Fig. 6.4. Makeshift houses made primarily of bamboo in the Mandla area of Dirang (Arunachal Pradesh) (photo: Baharul I. Choudhury).

Fig. 6.5. An adult female mithun (*Bos frontalis*) in Arunachal Pradesh (photo: Baharul I. Choudhury).

Fig. 6.6. *Shorea* forest in Assam (photo: Atiqur R. Barbhuiya).

vegetation classes such as sal, hollong, teak and chirpine forests. Bamboo occupies a very large area in the hill districts and is the primary vegetation of the recently abandoned 'jhum' cultivation area (Anonymous, 2002).

The state is home to a large number of endemic species such as *Shorea assamica* (Fig. 6.6), *Ilex khasiana*, *Bridelia assamica*, *Dendrobium farmari* and *Echinocarpus assamicus*. A few important medicinal plants are *Alstonia scholaris*, *Zanthoxylum nitidum*, *Andrographis paniculata*, *Oroxylum indicum*, *Chlerodendrum indicum*, *Jatropha curcas*, *Costus speciosus*, *Melastoma malabathricum*, *Aegle marmelos*, *Alpinia galangal*, *Artemisia indica*, *Centella asiatica*, *Garcinia cowa*, *Hedyotis scandens*, *Pogostemon pubescens*, *Pothos scandens*, etc. Major timber-yielding species of the state are *Dipterocarpus macrocarpus*, *Shorea assamica*, *Shorea robusta*, *Pinus* sp., *Schima wallichii* and *Tectona grandis*. Bamboo constitutes major vegetation in the state and supports a key economic return for the people. However, the state is under tremendous pressure from anthropogenic activities, especially due expansion of developmental projects (Fig. 6.7).

6.4.3 Manipur

Manipur is a small, hilly state (covering about 22,327 sq. km) just north of the Tropic of Cancer. It lies between 23°50' and 25°41' N latitudes and 93°6' and 94°47' E longitudes. The state is bounded on the north by Nagaland, on the south by Mizoram and the Chin Hills of Myanmar, on the east by the Chindwin districts of Myanmar and on the west by Cachar and North Cachar districts of Assam. Annual rainfall in the state ranges from 1,600–3,430 mm.

Clarke (1889) described 1,050 species of flowering plants in the state and ferns spread over 533 genera in 114 families and found a strong affinity to Khasia and Sikkim flora. Some of the largest genera are *Dendrobium* (43 spp.), *Polygonum* (23 spp.), *Impatiens* (17 spp. and 4 varieties), *Ficus* (18 spp.), *Coelogyne* (16 spp.), *Carax* (15 spp.), *Crotalaria* (14 spp.), and *Cyperus*, *Cymbidium* and *Jasminum*

Fig. 6.7. Forest fragmentation due to developmental activities (e.g. road construction) in Assam (photo: Lal Bihari Singha).

(13 spp. each). In a recent study 2,380 species belonging to 1,052 genera spread over 205 families of plants have been recorded from the state (Singh *et al.*, 2000).

Singh *et al.* (2000) classified the forests of Manipur into six types: (1) tropical moist deciduous forests; (2) tropical evergreen and semievergreen forests; (3) subtropical mixed forests; (4) subtropical pine forests; (5) subtropical and temperate grasslands; and (6) temperate forests. Aquatic ecosystems in the state are primarily represented by Loktak Lake (Ramsar site). There is enormous floral and faunal diversity here, including the endangered Manipuri brow-antlered deer (*Cervus eldi eldi*) (Gee, 1960). Loktak Lake also provides valuable ecosystem services to the local people (Fig. 6.8). Manipur is home to several primitive and endemic plant species, represented in Tables 6.4 and 6.5 respectively.

Singh *et al.* (2000) also reported many taxa which are facing different degrees of threat. These are *Acer laevigatum* Wallich, *Aconitum elwesii* Stapf, *Aconitum nagarum* Stapf, *Alangium* spp. *Annesklea fragrans* Wallich, *Anoectochilus tetrapterus* Hook. f. *Aquilaria malaccensis* Lam., *Aldrovanda vasiculosa* L. *Ascocentrum ampullaceum* Lindley ex Wallich, *Ascocentrum miniatum* (Lindley) Schltr., *Begonia obverse* C.B. Clarke, *Canarium strictum* Roxb., *Clematis apiculata* Hook. F. & Thomson, *Dalhausia bracteata* Grah., *Dendrobium bensonae* Reichb. f., *Dendrobium draconis* Reichb. f., *Dendrobium heterocarpum* Wallich ex Lindley, *Dendrobium wardianum* Warner and *Epigeneium amphum* (Lindley).

Bamboos along with other other interesting and economically important grasses are abundant in Manipur. Out of around 126 taxa of bamboos reported from India, over 41 species (*Arundinaria* ca 7 spp., *Bambusa* ca 15 spp., *Chmnobambusa* 1 sp., *Dendroclamus* ca 8 spp., *Dinochloa* 1 sp., *Melocanna* 1 sp., *Schizostachyum ca* 7 spp. and

Fig. 6.8. Local people fishing and harvesting minor produce from Loktak Lake in Manipur (photo: Lal Bihari Singha).

Thyrostachys 1 sp.) come from Manipur. The state also harbours several primitive cultivars and landraces of rice, maize, coix, etc. Wild relatives of cultivated plants that occur in the state are *Alpinia*, *Alocasia*, *Amomum*, *Mucuna*, *Pyrus*, *Prunus*, *Rubus* and *Sorbus*, etc. Other than these, eight species of *Dioscorea* and 21 species of piper are also found in Manipur. Wild species of banana such as *Ensete glaucum* (Roxb.) Cheesm., *Musa cheesmanii* Simmonds, *Musa nagensium* Prain, *Musa nagensium* Prain, *Musa balbisiana* Colla, palaeoendemic species such as *Gingko biloba* L. and neoendemic species of Orchid and *Primula*, etc. are also reported from the state.

6.4.4 Meghalaya

Meghalaya is a hill state of northeast India and is home to several interesting plant and animal species. The state lies between 25°00' and 26°01' N latitudes and 89°45' and 92°45' E longitudes with a geographical area of approximately 22,429 sq. km. Cherrapunji area in the state of Meghalaya receives rainfall throughout the year due to its orographic features and is one of the wettest places on earth, receiving as much as 11,246 mm rainfall per year (Guhathakurta *et al.*, 2015). Several primitive and endemic species occur in the state. The state is also considered to be the centre of origin for *Citrus* (Chatterjee *et al.*, 2006). Populations of *Citrus indica*, which is a wild relative of cultivated *Citrus*, are available in Nokrek Biosphere Reserve, Garo Hills. A few endemic species of Meghalaya are *Acranthera tomentosa*, *Adinandra griffithii*, *Begonia rubrovenia*, *Clematis apiculata*, *Crotalaria noveoides*, *Cyclea debiliflora*, *Dendroglossa minutula*, *Elaeagnus conferta*, *Inula kalapani*, *Neanotis oxyphylla*, *Pyrenaria khasiana*, *Sterculia khasiana* and *Trivalvaria kanjilali*. Habitat degradation due

Table 6.4. Primitive angiosperms of Manipur (adapted from Singh *et al.*, 2000).

Sl no.	Species	Family	Distribution
1	*Actinodephne angustifolia* Nees	Lauraceae	NE India/Malaya
2	*Actinodephne obov*ata (Nees) Blume	Lauraceae	NE India, eastern Himalaya and Bangladesh
3	*Alseodaphne sikkimnensis* Meissn	Lauraceae	NE India, Sikkim
4	*Alseodaphne owdenii* Parker	Lauraceae	NE India, Sikkim and Bhutan
5	*Artabotrys cubittii* Chaterjee	Annonaceae	Manipur Assam, Meghalaya and Myanmar
6	*Artabotrys hexapetalus* (L.f.) Bhandari	Annonaceae	NE India, Myanmar, Malesia
7	*Beilschmiedia assamica* Meissn.	Lauraceae	NE India, Manipur and Myanmar
8	*Beilschmiedia fagifolia* Nees	Lauraceae	NE India, Manipur
9	*Beilschmiedia roxburghiana* Nees	Lauraceae	NE India, Manipur
10	*Betula alnoides* Buch.-Ham	Betulaceae	Himalaya, NE India, Manipur
11	*Chloranthus elatior* Link	Chloranthaceae	NE India, Manipur, Himalaya and Indo-Malaya
12	*Chloranthus glaber* (Thunb.) Makino	Chloranthaceae	NE India and Indo-Malaya
13	*Corylopsis himalayana* Griffith ver. himalayana	Hamamelidaceae	NE India and Indo-China
14	*Cryptocarya amygdalina* Nees	Lauraceae	NE India, E. Himalaya and Andaman
15	*Desmos longoflorus* (Roxb.) Saff.	Annonaceae	NE India and Indo-Malaya
16	*Exbucklandia populnea* (R. Br. Ex Griffith) R.W. Brown	Hamamelidaceae	NE India, E. Himalaya and Indonesia
17	*Fissistigma bicolor* (Roxb.) Merr.	Annonaceae	NE India and west Bengal, Myanmar
18	*Fissistigma polyanthum* (Hook.f. & Thomson) Merr.	Annonaceae	NE India, Bhutan and Bangladesh
19	*Goniothalamus sesquipedalis* (Wallich) Hook. F. & Thomson	Annonaceae	NE India, Indo-Malaya and Myanmar

Continued

Table 6.4. Continued.

Sl no.	Species	Family	Distribution
20	*Holboellia latifolia* Wallich	Lardizabalaceae	NE India
21	*Horsfieldia amygadalina* (Wallich) Warb	Myristicaceae	NE India, Bangladesh and Myanmar
22	*Illicium griffithii* Hook. F & Thomson	Illiaceae	NE India, Eastern Himalaya
23	*Illicium manipurense* G. Watt ex King	Illiaceae	Endemic to Manipur
24	*Illicium simonsii* Maxim.	Illiaceae	NE India, eastern Himalaya and China
25	*Kadsura heteroclite* (Roxb.) Craib	Schizandraceae	NE India, eastern Himalaya and China
26	*Kadsura watti* C.B. Clarke	Schizandraceae	Endemic to Manipur
27	*Knema tenuinervia* Wilde.	Myristicaceae	NE India
28	*Knema linifolia* (Roxb.) Warb.	Myristicaceae	NE India, Bangladesh, Myanmar and Malaysia
29	*Knema tenuinervia* Wilde.	Myristicaceae	NE India, eastern Himalaya, Myanmar and Thailand
30	*Magnolia campbellii* Hook. f. & Thomson	Magnoliaceae	NE India, eastern Himalaya and China
31	*Magnolia griffithii* Hook. f. & Thomson	Magnoliaceae	Endemic to NE India
32	*Magnolia hodgsonii* (Hook. f. Thomson	Magnoliaceae	NE India, eastern Himalaya and Bangladesh
33	*Magnolia hookeri* (Cubitt & Smith) Rju & Nayar	Magnoliaceae	NE India and Myanmar
34	*Magnolia insignis* Wallich	Magnoliaceae	NE India, eastern Himalaya
35	*Michelia cathcartii* Hook. f. & Thomson	Magnoliaceae	Manipur, Meghalaya, Nagaland and Sikkim
36	*Michelia doltsopa* Buch.-Ham ex DC	Magnoliaceae	NE India, eastern Himalaya
37	*Michelia glabra* Parm.	Magnoliaceae	Endemic to NE India and eastern Himalaya
38	*Michelia punduana* Hook. f. & Thomson	Magnoliaceae	Endemic to NE India and eastern Himalaya
39	*Michelia velutins* DC.	Magnoliaceae	NE India, West Bengal and Nepal
40	*Polyalthia jenkinsii* Benth. Ex Hook. f.	Annonaceae	NE India and Indo-Malaya
41	*Tetracera sarmentosa* (L.) Vahl ssp. *anadamanica* (Hoogl.) Hoogl.	Dilleniaceae	NE India, West Bengal, Andamans, Myanmar and Malaysia

Table 6.5. Endemic plants of Manipur (Singh *et al.*, 2000).

Species	Family	Distribution
Aconitum elwesii Stapf	Rananculaceae	Manipur, Nagaland and Arunachal Pradesh
Aconitum nagarum Stapf	Rananculaceae	Manipur, Nagaland and Arunachal Pradesh
Actinodaphne obovata Blume var. *wattii* King	Lauraceae	Manipur
Agapetes lobbii C.B. Clarke	Vacciniaceae	Manipur
Agapetes mannii Hemsl	Vacciniaceae	Manipur, Meghalaya NE India
Arisaema wattii Hook. f.	Araceae	Manipur and NE India
Artabotrys cubitti Chatterjee	Annonaceae	NE India and Myanmar
Aster ageratoides Turez.	Asteraceae	Manipur
Asystasia pusilla C.B. Clarke	Acanthaceae	Manipur and Nagaland
Baliospermum suffruticosum Pax & Hoffm.	Euphorbiaceae	Manipur
Beaumonita longiotuba Craib	Apocyanaceae	Manipur
Begonia obverse C.B. Clarke	Begoniaceae	Manipur
Berberis manipurana Ahrendt	Berberidaceae	Manipur
Berberis sublevis W.W. Smith var. *sublevis*	Berberidaceae	Manipur
Carex manipurensis C.B. Clarke	Cyperaceae	Manipur
Clemetis apiculata Hook. f. & Thomson	Rananculaceae	Manipur and Meghalaya
Clemetis watti J.R. Drumm & Craib	Rananculaceae	Manipur
Colquhounia elegans Wallich ex Benth.	Lamiaceae	Manipur
Craibiodendron stellatum (Pierre) W.W. Smith	Ericaceae	Manipur
Dalbargia wattii C.B. Clarke	Fabaceae	Manipur
Dischidia micholitzii N.E. Brown	Asclepiadaceae	Manipur
Elatostema ciliatum C.B. Clarke	Urticaceae	Manipur
Elaeagnus loureirii Champion	Elaeagnaceae	Manipur
Epipremnum meeboldii Krause	Araceae	Manipur
Euphorbia serrulata Reinw. Ex Blume	Euphorbiaceae	Manipur
Garcinia loniceroides T. Anderson	Clusiaceae	Manipur & Nagaland
Gleadovia banarjiana Deb	Orobancaceae	Manipur
Hedychium greenii var. *urophyllum* W.W. Smith	Zingiberaceae	Manipur
Hedychium speciosum Wallich ex. Roxb.	Zingiberaceae	NE India
Hedychium wardii Fisher	Zingiberaceae	NE India
Ilex embelioides Hook. f. & Thomson	Aquifoliaceae	Manipur and Meghalaya

Continued

Table 6.5. Continued.

Species	Family	Distribution
Ilex khasiana Purkeay	Aquifoliaceae	NE India
Ilex wattii Loes	Aquifoliaceae	Manipur
Illicium manipurense G. Watt ex Roxb	Illiciacea	Manipur
Impatiens gibbisepala Hook. f.	Balsaminaceae	Manipur
Impatiens longirama Hook. f.	Balsaminaceae	Manipur
Impatiens rubro-lineata Hook. f.	Balsaminaceae	Manipur
Impatiens spissiflora Hook. f.	Balsaminaceae	Manipur
Impatiens teneriflora Hook. f.	Balsaminaceae	Manipur
Iris bakeri Wallich	Iridaceae	Manipur
Iris wattii Baker	Iridaceae	Manipur
Justicia anfructuosa C.B. Clarke	Acanthaceae	Manipur
Kadsura wattii C.B. Clarke	Schizandraceae	Manipur
Kalanchoe rosea C.B. Clarke	Crassulaceae	Manipur
Mahonia feddei Ahrendt	Berberidaceae	Manipur
Mahonia magnificia Ahrendt	Berberidaceae	Manipur
Mahonia manipurensis Takeda	Berberidaceae	Manipur
Mantisia saltatoria Sims	Zingiberaceae	Manipur and Mizoram
Musa itinerans Cheesman	Musaceae	Manipur
Phacellaria compressa Benth	Santalaceae	Manipur and Arunachal Pradesh
Pilea minuta C.B. Clarke	Utricaceae	Manipur
Pimpinella flaccida C.B. Clarke	Apiaceae	Manipur and Nagaland
Piper aurorubrum C. DC.	Piperaceae	Manipur
Piper gamblei C.B. Clarke	Piperaceae	Manipur, Mizoram and Nagaland
Piper kapruannum C. DC.	Piperaceae	Manipur
Piper lainatakanum C. DC.	Piperaceae	Manipur
Piper makruense C. DC.	Piperaceae	Manipur
Piper meeboldii C. DC.	Piperaceae	Manipur
Piper muneyporensis C. DC.	Piperaceae	Manipur
Piper nagense C. DC.	Piperaceae	Manipur and Nagaland
Polygonum stellato-tomentosum Smith & Ramaswami	Polygonaceae	Manipur
Potentilla manipurensis G. Watt	Rosaceae	Manipur
Prunus wattii Ghora and Panigr.	Rosaceae	Manipur
Pyrus wattii (Koehme) Bennet	Rosaceae	Manipur
Rosa clinopylla Thory var. *purvifolia* Ghora and Panigr.	Rosaceae	Manipur
Sorbus polycarpa (Hook. F.) Rehder	Rosaceae	Manipur and Meghalaya
Spodiopogon lacei (Torr.) Torr.	Poaceae	Manipur
Strychnos mux-blunda A.W. Hill	Loganiaceae	Manipur
Swertia wattii C.B. Clarke	Gentinaceae	Manipur and Nagaland

Continued

Table 6.5. Continued.

Species	Family	Distribution
Synotis rhabdos C.B. Clarke Jeffery & Chen	Asteraceae	Manipur and Nagaland
Vaccinium lamellatum Stevens	Vacciniaceae	Manipur
Vaccinium manipurensis G. Watt ex Brandis) Sleumer	Vacciniaceae	Manipur and Arunachal Pradesh
Vernonia clivorum Hance	Asteraceae	Manipur
Vernonia cylindriceps C.B. Clarke	Asteraceae	Manipur and Nagaland
Zanthoxylum pseudoxyphyllum Babu	Rutaceae	Manipur

to clear felling for shifting agriculture is prominent in the region, including protected areas (Fig. 6.9) causing significant damage to the pristine ecosystems.

6.4.5 Mizoram

Mizoram is the extreme southern fringe state of northeast India covering an area of 21,081 sq. km. The state is located in between 21°57'–24°30' N latitude and 92°15'–93°29' E longitude. Mizoram also experiences the presence of the Tropic of Cancer as it goes through the state. It is surrounded in the north by the Cachar District of Assam, in the east by the state of Manipur, in the east, south and southwest by the Chin and Arakan hills of Myanmar and on the western side by the state of Tripura and Chittagong hill tracts of Bangladesh. The state receives annual rainfall of between 2,000–3,600 mm from both northeast and southwest monsoons. In general, the forest is categorized into six types (tropical wet evergreen and semievergreen forests; montane subtropical forests; temperate forests; bamboo forests; quercus forests; and jhum land (Singh *et al.*, 2002). About 20 wild relatives of cultivated plants also originate in the state (Singh *et al.*, 2002).

Mizoram is home to 2,358 species of plants of which 2,141 (905 genera and 176 families) belong to angiosperms. Six species (six genera and four families) belong to gymnosperms and 211 species (66 genera and 35 families) belong to pteridophytes. Out of 2,141 angiosperms, 1,641 are dicots and the rest belong to monocots (Singh *et al.*, 2002). Major threats to the flora and fauna of Mizoram are agricultural activities, construction of new houses for settlement in forest areas, illegal encroachments into forest land, construction of new roads, urbanization, jhum cultivation, etc. Plant diversity in the homegardens (Fig. 6.10) of Mizoram is also enormous, where local people cultivate a wide range of plant species for their regular consumption. A recent study demonstrated occurrence of numerous plant species having an array of ethnobotanical uses (Barbhuiya *et al.*, 2016b).

Fig. 6.9. Clear felling of primary forest in Nokrek Biosphere Reserve in Meghalaya causing serious threat to the pristine ecosystem of the region (photo: Baharul I. Choudhury).

Fig. 6.10. Overview of a homegarden in Mizoram showing a wealth of species for multiple uses (photo: Atiqur R. Barbhuiya).

6.4.6 Nagaland

Nagaland is another hilly state located in the northeastern part of India and lies in between 25°10' and 27°01'N latitude and 93°17' and 95°15'E longitude. The state of Nagaland is also known to be the northernmost tropical forest formation of the earth (Chatterjee *et al.*, 2006). The geographical area is 16,600 sq. km and divided into seven districts: Kohima, Mokokchung, Mon, Phek, Tuensang, Wokha and Zunheboto. Physiographically, almost all of Nagaland is hilly and mountainous. The main hill ranges are the Barail in the south and southwest and Patkai in the north. A high range exists along the border with Myanmar and Mt Saramati (3,842 m) is the highest point in the range. The hills in the central areas are commonly referred to as the 'Naga Hills'. The highest peak of the Barail Range is Mt Japfu (Japvo) which stands at 3,043 m. Small plains occur along the Dhansiri river, especially near Dimapur, and to some extent near other rivers where they enter Assam. There are smaller mountain valleys in different areas. The climate of Nagaland is tropical 'monsoon' type, with a hot wet summer and a cool dry winter. Winter rains are also not uncommon.

The state is very rich in floral diversity and around 2,250 species have been reported so far (Singh, 2010). Bamboo resources are also well known (Fig. 6.11)

Fig. 6.11. A bamboo forest in Nagaland showing luxuriant growth in the natural habitat (photo: Lal Bihari Singha).

which are variously utilized for personal and commercial purposes. More than 70% of the population in Nagaland are dependent on a traditional agrarian economy and cultivate and maintain a wide variety of crops such as rice, millet, maize, pulses, sugarcane, potato, coffee, cardamom, tea and others (Singh, 2010).

6.4.7 Sikkim

This small hilly state with an area of 7,300 sq. km is situated on part of the eastern Himalaya and lies between 27°10' and 28°5' N latitude and 88°30' and 89°0' E longitude, with an average annual rainfall of around 4,000 mm. The state consists of four districts (East, West, North and South Sikkim) and is bounded by Nepal in the west, Bhutan in the southeast, Tibet in the north and north-east and Darjeeling district of West Bengal in the south. The altitude of the state ranges from 244 m to 8,598 m. Srivastava (1996) estimated the occurrence of 5,000 species of flowering plants, 350 pteridophytes and 15 gymnosperms from the state. Major forest types depending on elevation and characteristics of vegetation are tropical, temperate and alpine. Srivastava (1996) also enumerated Sikkim Himalayan region as a suitable spot for migration of floristic elements from several near and far-off lands. He also mentioned the occurrence of many such elements in the region (presented in Table 6.6). Other primitive genera and species are *Magnolia*, *Manglietia*, *Euptelea*, *Tetracentron*, *Pycnarrhena*, *Haematocarpus*, *Aspidocarya*, *Holboellia*, *Exbucklandia*, *Houttuyanta*, *Myrica*, *Alnus* and *Betula*. Hajra

Table 6.6. Floristic elements of Sikkim region from several near and far-off lands (adapted from Srivastava, 1996).

Floristic element	Species
Sino-Japanese element	*Quercus*, *Schima*, etc.
Western China element	*Aletris pauciflora*, *Anemone rupicola*, *Anemone vitifolia*, *Primula sikkimensis*, *Magnolia campbellii*, etc.
European and Mediterranean element	*Ranunculus*, *Gentiana*, *Swertia*, *Anemone*, *Tamarix*, *Allium*, *Artemisia*, etc.
American element	*Eupatorium adenophorum*, *Eupatorium odoratum*, *Parthenium hysterophorus*, *Mikania cordata* and *Lantana camara*, etc.
African element	*Flacourtia*, *Grewia*, *Holarrhena*, *Alstonia*
Siberian element	*Potentilla*, *Pedicularis* and *Lonicera*
Tibetan element	*Hippobhae*, *Prezwalskia*, etc.
Northern temperate and arctic region	*Glaux maritima*
Other isolation belt plants	*Meconopsis bella*, *Cathcartia lyrata*, *Senecio chola*, *Sausurea laneana*, *Geranium* spp., *Primula elwesiana*, *Primula wattii and Swertia burkilliana*

and Verma (1996) reported several endangered taxa of Sikkim which are facing different degrees of natural and anthropogenic pressure. A few such plant species are *Acer hookeri* var. *majus*, *Aconitum ferox*, *Acronema pseudotonera*, *Angelica bulbigena*, *Aphyllorchis parviflora*, *Begonia rupicola*, *B. sanctata*, *B. strapis*, *Cymbidium eburneum*, *C. hookerianum*, *C. whiteae*, *Ceropegia hookeri*, *C. lucida*, *Calamus inermis*, *Calanthe alpine*, *Carex kingiana*, *Christiopteris tripcuspis*, *Codonopsis affinis*, *Coelogyne treutleri*, *Cotoneaster simonsii*, *Cyclogramma squanaestipes*, *Cypripedium himalaicum*, *Dennataedtia elwesii*, *Didiciea cunninghammii*, *Diplomeris hirsute*, *Lagerstroemia minuticarpa*, *Livistonia jenkisiana*, *Lloydia himalensis*, *Macodium levingei*, *Nardostachys grandiflora*, *Ophiorrhiza lucida*, *Oreopteris elwesii*, *Panax pseudoginseng*, *Paphipedilum venusatum*, *Picrorrhiza kurrooa*, *Pimpinella tongloensis*, *P. wallichii*, *Rhopalocnomis phalloides*, *Rhynchospora sikkimensis*, *Vanda spectabilis* and *Zeuxine pulchra*.

Hajra and Verma (1996) also reported a few endemic plants in the Sikkim Himalayan region. A few endemics of Sikkim are *Anaphalis cavei*, *A. hookeri*, *A. subumbellata*, *Agrostis neodebilis*, *Artemisia thellungiana*, *Blumea sikkimensis*, *Cremanthodium decaisnei* f. *clarkei*, *C. palpatum* ssp. *benthamii*, *Calamagrostis tripilifera* var. *tripilifera*, *C. tripilifera* var. *cimminsit*, *Carex kingiana*, *Catabrosa aquatica*, *Coelogyne treutleri*, *Crepis atropappa*, *Cyathopus sikkimensis*, *Drepanostachyum intermedium*, *Gentiana glabriuscula*, *G. pluviarum*, *G. prainii*, *G. recurvata*, *Inula macrosperma*, *Jaeschkea microsperma*, *J. smithii*, *Ligularia dur*, *L. hookeri* ssp. *clarkei*, *L. pachycarpa*, *Poa gammieana*, *Rhynchospora sikkimensis*, *Saussurea forrestii*, *S. lineana*, *S. nimborum*, *S. obscura*, *S. pantlingiana*, *S. rex*, *Swertia ramose*, *Trisetum sikkimense*, etc.

6.4.8 Tripura

This easternmost state of northeast India is bordered by international boundaries of Bangladesh to the west, south and north, Assam to the northeast and Mizoram to the east. The area of the state is 10,477 sq. km and lies between 22°56' and 24°32' N latitudes and 91°10' and 92°21' E longitudes. The state has a tropical monsoon climate with peak rainy season from June to September. The average annual precipitation of the state is around 200 mm. According to a recent report (Anonymous, 2002), the Tripura forest areas support most of the tree species of broad-leaved and deciduous semievergreen forests to a significant extent, confined to the less hilly areas in the west and southwestern part of the state. Dominant species of such forest areas are *Shorea robusta*, *Schima wallichii*, *Artocarpus chama*, *Careya arborea*, *Dillenia pentagyna*, *Cassia fistula*, *Albizia procera*, etc. About 14% of species found in Tripura are endemic (Chatterjee *et al.*, 2006).

6.5 Future Conservation Initiatives in Northeast India

It has been noted from the present observation that notheast India is home to several primitive, endemic, rare, endangered and threatened species of ecological and economic significance. However, such an array of natural resources is on the verge of erosion due to various natural and anthropogenic disturbances. Moreover, such disturbances are also prominent even in protected areas such as national parks and sanctuaries and pose a serious threat to the native flora and fauna of the region. Therefore, the concern for environmental degradation and side-by-side conservation of such unparalleled resources has become imperative for the region.

In general, an adequate conservation plan for an endangered species requires that as many individuals as possible be preserved within the greatest possible area or habitat. However, this does not provide an effective species protection strategy. Several measures have been recommended for the effective conservation of important species throughout the world. Restoration and conservation of habitat as well as individual species have become a mission-oriented science aimed at preserving biological diversity. Shaffer (1981) defined the number of individuals needed to ensure the necessary survival of a species as its minimum viable population (MVP). An MVP is the smallest population that can be predicted to have a very high chance of persisting for the foreseeable future. In the case of small populations which are subject to rapid decline in numbers and are in real danger of extinction, establishment of new populations will be effective in enriching the natural population.

A few important conceptual issues pertaining to restoration ecology range from sociopolitical, including organization and management of restoration efforts, to biological, such as population viability analysis, conservation genetics, metapopulation biology, landscape ecology, etc. (Bowles and Whelan, 1994). Reintroduction of threatened species to their natural habitat is also an effective tool for increasing the natural population of such species. However, such restoration of endangered species depends on the biology of the organism and, for that, knowledge of life history of the target species is more important. Since rare and endangered taxa often exist as relatively small populations (Holsinger and Gottlieb, 1989), they may have been subject to past population bottlenecks and genetic manipulation may be required to recover declining populations or to restore or maintain evolutionary potential (Fenster and Dudash, 1994). Otherwise, there will be a rapid loss of genetic variability as well as alleles from such small populations (Frankham, 1996). Thus, a strategic approach towards effective conservation of resources will be beneficial for the successful conservation of the rich bioresources of the region.

6.6 Effective Conservation Strategies for *G. assamicus*

Conservation of rare/endangered species is the biggest challenge of the century. Scientific communities and policy-makers now broadly accept the importance of biological diversity; however the achievement of such still remains uncertain. Overharvesting of plants for food, medicines, timber and similar uses often leads to rapid loss of plant genetic resources and ultimately to extinction. There are several approaches available for conserving 'species at risk' in both in situ and ex situ conditions.

6.6.1 Conservation through cultivation

'Conservation through cultivation' (CTC) is one of the successful approaches for management of different plant species. A CTC programme was developed by Limbe Botanic Garden in Mount Cameroon to conserve the rich and fragile biodiversity through reducing harvesting pressures and providing cultivated material (Nkefor *et al.*, 1999). The success story of *Prunus africana* in Limbe Botanic Garden (Sunderland and Nkefor, 1997) showed a pathway for conservation for similar species at risk. CTC initiatives gained more popularity after their initial success and have been applied to several species of ecologic and economic importance such as the Brazil nut (*Bertholletia excelsa*) (Paiva *et al.*, 2011) and bitter mango *(Irvingia wombolu)* (Asaah *et al.*, 2003). There are several ongoing CTC initiatives for similar species such as American ginseng (*Panax quinquefolius*) (Anonymous, 2018), and many orchid species have also been recovered from the verge of extinction through cultivation for their economic and aesthetic values.

The other most successful and popular CTC is the story of the oldest tree on earth known as the 'living fossil' (*Ginkgo biloba*). *Ginkgo biloba* remained unchanged for more than 200 million years. However, the species has been declared extinct in the wild for centuries. The ginkgo tree is adored in many parts of the world as a street tree and ornamental tree, for its beauty and longevity, as well as for its medicinal values. It was only human intervention and a CTC approach that ensured survival of ginkgo in a wide geographic area (Cohn, 2013). There are many reasons for successful implementation of the CTC approach for *G. assamicus*. Firstly, highly saponaceous pods are used for multiple purposes (Choudhury *et al.*, 2007). Mature pods are preserved by the local people and offered as a precious gift to the Buddhist monks, relatives and friends. Therefore, cultivation in homegardens is an ideal approach to augment the population locally. Secondly, *G. assamicus* leaves turn an elegant, bright yellow colour during autumn (September to November) and are distinctly visible from a far distance (see cover photo). Trees with such attractive appearance are planted as ornamentals along streets, in parks and in botanical gardens. *Gleditsia triacanthos*, a sister genus of *Gymnocladus*, is widely planted as a street tree throughout North America for its fall colour and physiognomy and is a popular tree throughout the region (Fig. 6.12). Kentucky coffeetree (*Gymnocladus dioica*) is also planted along roadsides across cities in North America (Fig. 6.13).

G. assamicus is a 15–20 m tree with moderate spread and upright silhouette. Light green young leaves provide filtered shade and turn bright yellow at the later stage. Though the purple coloured flowers are not 'blazing', mature pods can be used as soap substitute in rural areas of Arunachal Pradesh and adjoining areas in northeast India. Local people also mentioned that mature pods are highly favoured by deer and wild boar. Therefore, an increase in *G. assamicus* population in homegardens, on agricultural land, in pastures and at roadsides in mountainous terrain of the region will be helpful for the survival and growth of diminishing wildlife in the region.

Fig. 6.12. A closely related species (*Gleditsia tricanthos*) of *Gymnocladus assamicus* is planted along roadsides in Montreal (Canada) (photo: Baharul I. Choudhury).

6.6.2 Silvicultural approaches and protected area management

Population restoration by supplementary regeneration is an adaptive strategy to combat the declining population of *G. assamicus*. Silvicultural intensification is one of the best methods of conservation for species which are overharvested from their natural habitat (Fredericksen and Putz, 2003). Sparse populations of *G. assamicus* are scattered across a wide geographic area (Menon *et al.*, 2010). Forest areas in the locality are generally community-owned which gives extensive control over the forest resources for local utilization. In situ conservation strategies by declaring designated areas as protected is often challenging in such situations. However, attempts have been made to designate community conserved areas (CCAs) in the locality that show rich conservation potential (WWF, 2019).

6.6.3 Community-based conservation initiatives

Since major threats to *G. assamicus* are habitat degradation and overharvesting of mature pods, awareness among local people, preservation of existing reproducing individuals and sustainable harvest of mature pods may be effective in successful conservation of the species. One of the most efficient ways to achieve this would be by introducing nursery-grown seedlings to compensate

Fig. 6.13. A Kentucky coffeetree plantation along a roadside in Toronto (Canada) (photo: Baharul I. Choudhury).

for the lack of natural regeneration. As an initial step, local communities have been involved in conservation of the existing population and several awareness meetings have been conducted to preserve *G. assamicus* from the danger of extinction. As a part of the conservation initiative, nursery-grown seedlings were distributed to the local people for increasing the population of soap pod tree in its natural habitat (Fig. 6.14). Government agencies such as the State

Fig. 6.14. Soap pod seedling distribution among the local people to augment population in its native range (photo: Baharul I. Choudhury).

Forest Department, non-governmeantal organizations (NGOs) and other conservation agencies may take initiatives in developing and distributing as well as planting nursery-grown seedlings in suitable habitats. Homegardens play a significant role as storehouses of rare/endangered tree species (Das and Das, 2005). For this, plantation of seedlings in homegardens, along traditional agricultural land or at waysides is also recommended. Such locations are generally protected and therefore the growth and survival of seedlings may be ensured.

Sacred groves are important centres of biodiversity conservation in northeast India (Mgumia and Oba, 2003; Khumbongmayum *et al.*, 2005; Bhagwat and Rutte, 2006). Furthermore, in situ conservation initiatives have been initiated in association with the local community by declaring *G. assamicus* as sacred in the community. To achieve this, mature trees were marked with traditional signs (Fig. 6.15) to protect them from felling, over-harvesting or any such damage. The local Monpa tribe follow certain community myths, norms and spirits that control over-harvesting of natural resources and are quite aware of biodiversity conservation and natural resource management. Therefore, such initiatives may have promising results for conservation of endangered species in the region.

6.6.4 In vitro method of conservation

A revolutionary technique of in vitro conservation of rare and endangered plants is promising and has been very effective so far. Various in vitro methods such as micropropagation, in vitro seed germination, dual culture with symbiotic fungi,

Fig. 6.15. Attempt to declare soap pod a sacred tree by putting signs on the tree trunk (highlighted with red circles) (photo: Baharul I. Choudhury).

and regeneration from callus are suitably applied based on availability of materials to conserve species of interest (Fay, 1992). In vitro micropropagation techniques have been optimized for *G. assamicus* using nodal explants by Gupta *et al.* (2013). In this approach, mature *G. assamicus* seeds were germinated and grown on soil media to obtain the nodal explants which were then cultured on Murashige and Skoog basal medium supplemented with 1.5 mgl-1 benzyl adenine and 50 mgl-1 adenine sulphate (Gupta *et al.*, 2013). Around 65% of the in vitro-raised plants were able to establish successfully in the greenhouse and field conditions (Gupta *et al.*, 2013). The Royal Botanic Gardens, Kew (UK) has established a major micropropagation unit and has an extensive collection of in vitro plants including many threatened species from throughout the world (Sarasan *et al.*, 2006). Further to this, we deposited seeds at the NBPGR (India) for long-term ex situ conservation of germplasm in gene banks (accession no IC-551620).

6.7 Future Studies

Success of any conservation initiative depends on several factors including the genetic diversity of the organisms. In general, species facing extinction go through a severe genetic bottleneck, primarily due to decline in population size (Hedrick,

2001; Allendorf, 2017). Such species also experience erosion of genetic variation and increased interpopulation genetic divergence due to increased random genetic drift, elevated inbreeding and reduced gene flow (Young *et al.*, 1996). Population genetic analysis of *G. assamicus* is therefore extremely important to develop long-term conservation goals so that populations with the highest genetic diversity are protected or utilized for further regeneration. However, the 'critically endangered' status of the species is a major challenge in conducting such study.

Population viability analysis (PVA) is a process of evaluation of data and models for a population to anticipate the likelihood that a population will persist for some arbitrarily chosen time into the future (Shaffer, 1981, 1987). Such analysis has become a commonly used tool in endangered species management and has been widely applied in conservation biology to predict extinction risks for threatened species and to compare alternative options for their management. A few examples are Furbish's lousewort (*Pedicularis furbishiae*) (Menges, 1990), *Gentiana pneumonanthe* (Oostermeijer *et al.*, 2000) and American ginseng (*Panax quinquefolium*) (Nantel *et al.*, 1996), and animals such as the bay checkerspot butterfly (*Euphydryas editha bayensis*) (Murphy *et al.*, 1990), the African elephant (*Loxodonta africana*) (Armbruster and Lande, 1993) and the helmeted honeyeater (*Lichenostomus melanops cassidix*) (Akçakaya *et al.*, 1995). PVA is widely applied in conservation biology to predict extinction risks for threatened species and to compare alternative options for their management.

6.8 References

Akçakaya, H.R., McCarthy, M.A. and Pearce, J.L., 1995. Linking landscape data with population viability analysis: management options for the helmeted honeyeater *Lichenostomus melanops cassidix*. *Biological Conservation* 73(2): 169–176.

Allendorf, F.W., 2017. Genetics and the conservation of natural populations: allozymes to genomes. *Molecular Ecology* 26(2): 420–430.

Anonymous, 2002. *Biodiversity Characterisation at Landscape Level in North-East India using Satellite Remote Sensing and Geographic Information System.* Department of Biotechnology Government of India & Department of Space Government of India, Indian Institute of Remote Sensing, Department of Space, Government of India.

Anonymous, 2018. *Recovery Strategy for the American Ginseng (Panax quinquefolius) in Canada 2018,* https://www.canada.ca/en/environment-climate-change/services/species-risk-public-registry/recovery-strategies/american-ginseng-2018.html, accessed 20 December 2018.

Anonymous, 2019. *Plant Diversity of Assam.* ENVIS Centre: Assam, Status of Environment and Related Issues, http://asmenvis.nic.in/Database/Plant_Diversity_833.aspx, accessed 26 January 2019.

Armbruster, P. and Lande, R., 1993. A population viability analysis for African elephant (*Loxodonta africana*): how big should reserves be? *Conservation Biology* 7(3): 60–-610.

Asaah, E.K., Tchoundjeu, Z. and Atangana, A.R., 2003. Cultivation and conservation status of *Irvingia wombolu* in humid lowland forest of Cameroon. *Journal of Food Agriculture And Environment* 1: 251–256.

Barbhuiya, A.R., Khan, M.L. and Dayanandan, S., 2016a. Genetic structure and diversity of natural and domesticated populations of *Citrus medica L.* in the Eastern Himalayan region of Northeast India. *Ecology and Evolution* 6(12): 3898–3911.

Barbhuiya, A.R., Sahoo, U.K. and Upadhyaya, K., 2016b. Plant diversity in the indigenous home gardens in the Eastern Himalayan Region of Mizoram, Northeast India. *Economic Botany* 70(2): 115–131.

Bhagawati, B.C., 2002. Social structure in Arunachal Pradesh: An anthropological overview. In R.C. Sundriyal, T. Singh and G.N. Sinha (eds) *Arunachal Pradesh: Environmental Planning and Sustainable Development.* G.B. Pant Institute of Himalayan Environment & Development, Kosi-Katarmal, Almora, India.

Bhagwat, S.A. and Rutte, C., 2006. Sacred groves: potential for biodiversity management. *Frontiers in Ecology and the Environment* 4(10): 519–524.

Bhattacharyya, B., 1967. Tertiary plant fossils from Cherrapunji and Laitryngew in Khasi and Jaintia Hills, Assam. *Quarterly Journal of the Geological, Mining and Metallurgical Society of India* 39(2): 131–134.

Bhattacharyya, B., 1983. Fossil plants from the Tura formation (Eocene) in the Garo Hills, Meghalaya. *Indian Journal of Earth Science* 10(1): 1–10.

Bhattacharyya, B., 1985. Leguminous fruits from the Eocene of Garo Hills, Meghalaya. *Quarterley Journal of the Geological, Mining and Metallurgical Society of India* 57: 215–225.

Bose, M.N., and Sah, S.C.D., 1964. Plant remains from Laitryngew, Assam. *Palaeobotanist* 12: 220–223.

Bowles, M.L. and Whelan, C.J. 1994. Conceptual issues in restoration ecology. In M.L. Bowles and C.J. Whelan (eds) *Restoration of Endangered Species: Conceptual Issues, Planning and Implementation.* Cambridge University Press, Cambridge, pp. 34–62.

Champion, S.H. and Seth, S.K., 1968. *A Revised Survey of the Forest Types of India.* Manager of Publications, Delhi.

Chatterjee, S., Saikia, A., Dutta, P., Ghosh, D., Pangging, G. and Goswami, A.K., 2006. Biodiversity significance of North East India for the study on natural resources, water and environment nexus for development and growth in northeastern India. Forest Conservation Programme: New Delhi (India). *Background paper no. 13.* World Wildlife Fund. New Delhi.

Chitale, V.S., Behera, M.D. and Roy, P.S., 2014. Future of endemic flora of biodiversity hotspots in India *PloS one*, 9(12): e115264.

Chivian, E. and Bernstein, A. (eds), 2008. *Sustaining Life: How Human Health Depends on Biodiversity.* Oxford University Press, Oxford.

Choudhury, B.I., Khan, M.L., Arunachalam, A. and Arunachalam, K., 2007. *Gymnocladus assamicus* Kanjilal ex P.C. Kanjilal fruit – a soap substitute. *Natural Product Radiance* 6: 427–429.

Choudhury, B., Khan, M.L. and Dayanandan, S., 2013. Genetic structure and diversity of indigenous rice (*Oryza sativa*) varieties in the Eastern Himalayan region of Northeast India. *SpringerPlus* 2(1): 228.

Clarke, C.B., 1889. On the plants of Kohima and Muneypore. *Journal of the Linnean Society of London, Botany* 25(165–169): 1–107.

Cohn, R., 2013. Ginkgo: The life story of the oldest tree on earth. Interview published in *Yale Environment* 360, http://e360.yale.edu/feature/peter_crane_history_of_ginkgo_earths_oldest_tree/2646/, accessed on 27 January 2019.

Conservation International, 2006. *Biodiversity Hotspots*, https://web.archive.org/web/20120327075212/http://www.biodiversityhotspots.org/xp/hotspots/Documents/cihotspotmap.pdf, accessed 5 April 2019.

Conservation International, 2019. *What's a Hotspot?* https://www.conservation.org/priorities/biodiversity-hotspots, accessed 29 January 2019.

Das, T. and Das, A.K., 2005. Inventorying plant biodiversity in homegardens: A case study in Barak Valley, Assam, North East India. *Current Science* 89(1): 155–163.

Fay, M.F., 1992. Conservation of rare and endangered plants using in vitro methods. *In Vitro Cellular & Developmental Biology-Plant* 28(1): 1–4.

Fenster, C.B. and Dudash, M.R., 1994. Genetic considerations for plant population restoration and conservation. In M.L. Bowels and C.J. Whelan (eds) *Restoration of Endangered Species: Conceptual Issues, Planning and Implementation.* Cambridge University Press, Cambridge, pp. 34–62.

Frankham, R., 1996. Relationship of genetic variation to population size in wildlife. *Conservation Biology* 10(6): 1500–1508.

Fredericksen, T.S. and Putz, F.E., 2003. Silvicultural intensification for tropical forest conservation. *Biodiversity & Conservation* 12(7): 1445–1453.

Gajurel, P.R., 2002. Experimental taxonomic studies on *Piper* (Piperaceae) of Arunachal Pradesh. PhD thesis, North-eastern Hill University, Shillong, India.

Gee, E.P., 1960. Report on the status of brow-antlered deer of Manipur (India). *Journal of Bombay Natural History Society* 57(3): 597–617.

Government of Arunachal Pradesh, 2019. Official State portal, http://www.arunachalpradesh.gov.in/, accessed 5 April 2019.

Guhathakurta, P., Menon, P. and Nipane, N.B., 2015. Identifying the changes in rainfall pattern and heavy rainfall events during 1871–2010 over Cherrapunji. In K. Ray, M. Mohapatra, B.K. Bandyopadhyay and L.S. Rathore *High-Impact Weather Events over the SAARC Region.* Springer, Cham, Switzerland, pp. 335–345

Gupta, S., Kaliamoorthy, S., Mao, A.A. and Sarma, S., 2013. In vitro regeneration of *Gymnocladus assamicus* Kanjilal ex PC Kanjilal, an ethnomedicinal critically endangered and endemic tree species from Northeast India. *In vitro* 3(2): 204–213.

Hajra, P.K. and Verma, D.M., 1996. *Flora of Sikkim*, vol. I. *Monocotyledons: Flora of India Series – 2.* Botanical Survey of India, Kolkata, Ministry of Environment and Forests, Govt of India.

Hajra, P.K., Verma, D.M. and Giri, G.S., 1996. *Materials for the Flora of Arunachal Pradesh*, vol. I. Botanical Survey of India. Kolkata, Ministry of Environment and Forests, Govt of India, p. 401.

Hedrick, P.W., 2001. Conservation genetics: where are we now? *Trends in Ecology & Evolution* 16(11): 629–636.

Hegde, S.N., 2002. *Arunachal Pradesh State Biodiversity Action Plan.* Arunachal Pradesh Forest Research and Development Agency, State Forest Research Institute, Itanagar, Arunachal Pradesh, India.

Holsinger, K.E. and Gottlieb, L.D., 1989. The conservation of rare and endangered plants. *Trends in Ecology & Evolution* 4(7): 193–194.

IUCN, 2019a. The IUCN Red List of Threatened Species. Version 2019-1. https://www.iucnredlist.org/resources/summary-statistics#Summary%20Tables, accessed 11 June 2019.

IUCN, 2019b. International Union for the Conservation of Nature, Country Profile, India, https://www.iucn.org/asia/countries/india, accessed 29 January 2019.

Kaul, R.N. and Haridasan, K., 1987. Forest types of Arunachal Pradesh: a preliminary study. *Journal of Economic and Taxonomic Botany* 9(2): 379–389.

Khumbongmayum, A.D., Khan, M.L. and Tripathi, R.S., 2005. Sacred groves of Manipur, northeast India: biodiversity value, status and strategies for their conservation. *Biodiversity & Conservation 14*(7): 1541–1582.

Lakhanpal, R.N., 1954. Recognizable species of Tertiary plants from Damalgiri in the Garo Hills, Assam. *Palaeobotanist* 3: 27–31.

Lakhanpal, R.N., 1955. On the occurrence of Nelumbium in the Tertiary of Assam. *Journal of the Indian Botanical Society* 34: 222–224.

Lakhanpaul, S., Velayudhan, K.C. and Bhat, K.V., 2003. Analysis of genetic diversity in Indian taro [*Colocasia esculenta* (L.) Schott] using random amplified polymorphic DNA (RAPD) markers. *Genetic Resources and Crop Evolution* 50(6): 603–609.

Menges, E.S., 1990. Population viability analysis for an endangered plant. *Conservation Biology* 4(1): 52–62.

Menon, S., Choudhury, B.I., Khan, M.L. and Peterson, A.T., 2010. Ecological niche modeling and local knowledge predict new populations of *Gymnocladus assamicus* a critically endangered tree species. *Endangered Species Research* 11(2): 175–181.

Mgumia, F.H. and Oba, G., 2003. Potential role of sacred groves in biodiversity conservation in Tanzania. *Environmental Conservation* 30(3): 259–265.

Mittermeier, R.A., Turner, W.R., Larsen, F.W., Brooks, T.M. and Gascon, C., 2011. Global biodiversity conservation: the critical role of hotspots. In F.E. Zachos and J.C. Habel (eds) *Biodiversity Hotspots*. Springer, Berlin, Heidelberg, pp. 3–22.

Murphy, D.D., Freas, K.E. and Weiss, S.B., 1990. An environment-metapopulation approach to population viability analysis for a threatened invertebrate. *Conservation Biology* 4(1): 41–51.

Myers, N., 1988. Threatened biotas: 'hot spots' in tropical forests. *Environmentalist* 8(3): 187–208.

Nantel, P., Gagnon, D. and Nault, A., 1996. Population viability analysis of American ginseng and wild leek harvested in stochastic environments. *Conservation Biology* 10(2): 608–621.

Nkefor, J.P., Ndam, N., Blackmore, P.C. and Sunderland, T.C.H., 1999. In T.C.H. Sunderland, L.E. Clark and P. Vantomme (eds) *Current Research Issues and Prospects for Conservation and Development.* International Expert Meeting on Non-Wood Forest Products in Central Africa, The Limbe Botanic Garden, Cameroon held during 10–15 May, FAO, Rome, 1999.

Olson, D.M., Dinerstein, E., Wikramanayake, E.D., Burgess, N.D., Powell, G.V., Underwood, E.C., D'amico, J.A., Itoua, I., Strand, H.E., Morrison, J.C. and Loucks, C.J., 2001. Terrestrial ecoregions of the world: a new map of life on earth, a new global map of terrestrial ecoregions provides an innovative tool for conserving biodiversity. *BioScience 51*(11): 933–938.

Oostermeijer, J.G.B., Young, A. and Clarke, G., 2000. Population viability analysis of the rare *Gentiana pneumonanthe*: importance of genetics, demography, and reproductive biology. In A.G. Yang and G.M. Clarke (eds) *Genetics, Demography and Viability of Fragmented Populations*, pp. 313–334.

Paiva, P.M., Guedes, M.C. and Funi, C., 2011. Brazil nut conservation through shifting cultivation. *Forest Ecology and Management* 261(3): 508–514.

Paul, A., Khan, M.L., Arunachalam, A. and Arunachalam, K., 2005. Biodiversity and conservation of rhododendrons in Arunachal Pradesh in the Indo-Burma biodiversity hotspot. *Current Science* 89(4): 623–634.

Pawar, S., Koo, M.S., Kelley, C., Ahmed, M.F., Chaudhuri, S. and Sarkar, S., 2007. Conservation assessment and prioritization of areas in Northeast India: priorities for amphibians and reptiles. *Biological Conservation* 136(3), 346–361.

Poffenberger, M., Barik, S.K., Choudhury, D., Darlong, V., Gupta, V., Palit, S. and Upadhyay, S., 2006. Forest sector review of Northeast India. *Background paper.* 12.

Procter K.H., Haridasan, K. and Smith, G.W., 1998. How far does lowland tropical rainforest go? *Global Ecology and Biogeography* 7: 141–146.

Sajeev, S., Roy, A.R., Iangrai, B., Pattanayak, A. and Deka, B.C., 2011. Genetic diversity analysis in the traditional and improved ginger (*Zingiber officinale* Rosc.) clones cultivated in North-East India. *Scientia Horticulturae* 128(3): 182–188.

Sarasan, V., Cripps, R., Ramsay, M.M., Atherton, C., McMichen, M., Prendergast, G. and Rowntree, J.K., 2006. Conservation in vitro of threatened plants progress in the past decade. *In Vitro Cellular & Developmental Biology – Plant* 42(3): 206–214.

Shaffer, M.L., 1981. Minimum population sizes for species conservation. *BioScience*, 31(2): 131–134.
Shaffer, M., 1987. Minimum viable populations: coping with uncertainty. In M.E. Soute (ed.) *Viable Populations for Conservation*, Cambridge University Press, Cambridge, pp. 69–86.
Singh, A.K., 2010. Probable agricultural biodiversity heritage sites in India: VI. The northeastern hills of Nagaland, Manipur, Mizoram, and Tripura. *Asian Agri-History* 14(3): 217–243.
Singh, J.S. and Chaturvedi, R.K., 2017. Diversity of ecosystem types in India: a review. *Proceedings of the Indian National Science Academy–INSA* 83(3): 569–594.
Singh, N.P. Singh, D.K., Hajra, P.K. and Sharma, B.D., 2000. *Flora of India. Introductory Volume, 2*. Botanical survey of India. Ministry of Environment and Forests, Govt of India.
Singh, N.P., Singh, K.P. and Singh, D.K., 2002. *Flora of Mizoram, Volume I*. Botanical Survey of India, Kolkata, Ministry of Environment and Forests, Govt of India.
Singh, T. and Prakash, U., 1980. Leaf impressions from the Siwalik sediments of Arunachal Pradesh, India. *Geophytology* 10: 104–107.
Srivastava, R.C., 1996. Introduction In: Hajra, P.K. and Verma, D.M. (eds) *Flora of Sikkim, Vol. I. (Monocotyledons)*. Botanical Survey of India, Kolkata, Ministry of Environment and Forests, Govt of India.
State of Forest Report, 2017. *Forest Survey of India*. Ministry of Environment & Forests, Kaulagarh Road, P.O -IPE, Dehradun (India).
Sunderland, T. and Nkefor, J., 1997. Conservation through cultivation: a case study of the propagation of Prunus Africana. In A Strategy for the Conservation of *Prunus africana* on Mount Cameroon. Technical Papers and Workshop Proceedings. Tropical Agriculture Association. pp. 68–87.
Takhtajan, A.L., 1969. *Flowering Plants: Origin and Dispersal*. Smithsonian Institution Press, Washington, DC, p. 310.
Thomas, C.D., Williams, S.E., Cameron, A., Green, R.E., Bakkenes, M., Beaumont, L.J., Collingham, Y.C., Erasmus, B.F., De Siqueira, M.F., Grainger, A. and Hannah, L., 2004. Biodiversity conservation: uncertainty in predictions of extinction risk/effects of changes in climate and land use/climate change and extinction risk (reply). *Nature* 430(6995).
UNEP-WCMC, 2014. Biodiversity A–Z website: https://www.biodiversitya-z.org/. UNEP-WCMC, Cambridge, UK, accessed January 29, 2019.
Valdiya, K.S., 2002. Geodynamic perspective of Arunachal Pradesh: bearing on environmental security and planning for development. In R.C. Sundriyal, T. Singh and G.N. Sinha (eds) *Arunachal Pradesh: Environmental Planning and Sustainable Development – Opportunities and Challenges*. Bishen Singh Mahendra Pal Singh, Dehradun, India, pp. 1–15.
Vishnu-Mittre, 1969. Pollen stratigraphy of India. *Puratattva Bulletin of the Indian Archeological Society*, 13.
Wildlife Institute of India, 2007, http://www.wii.gov.in/biodiversity, accessed 12 August 2019.
Williams, K.J., Ford, A., Rosauer, D.F., De Silva, N., Mittermeier, R., Bruce, C., Larsen, F.W. and Margules, C., 2011. Forests of East Australia: the 35th biodiversity hotspot. *Biodiversity Hotspots*. Springer, Berlin, Heidelberg, pp. 295–310.
Wood, A., 2000. An emerging consensus on biodiversity loss. In A. Wood, P. Steduan-Edwards and J. Mang (eds) *The Root Causes of Biodiversity Loss*. Earthscan Publications, London, pp. 1–10.
WWF, 2019. *Cultures of Conservation*, https://www.wwfindia.org/news_facts/panda/cultures_of_conservation.cfm, accessed 21 January 2019.
Young, A., Boyle, T. and Brown, T., 1996. The population genetic consequences of habitat fragmentation for plants. *Trends in Ecology & Evolution* 11(10): 413–418.

Index